International Conference on Electronics, Engineering Physics and Earth Science (EEPES 2023)

E3S Web of Conferences Volume 404 (2023)

Kavala, Greece
21 - 23 June 2023

Editors:

Teodor Iliev
Ivaylo Stoyanov
Grigor Mihaylov
Jacob Fantidis

ISBN: 978-1-7138-8367-8

Printed from e-media with permission by:

Curran Associates, Inc.
57 Morehouse Lane
Red Hook, NY 12571

Some format issues inherent in the e-media version may also appear in this print version.

For additional information, please contact EDP Sciences – Web of Conferences
at the address below.

EDP Sciences – Web of Conferences
17, Avenue du Hoggar
Parc d'Activité de Courtabœuf
BP 112
F-91944 Les Ulis Cedex A
France

Phone: +33 (0) 1 69 18 75 75
Fax: +33 (0) 1 69 28 84 91

contact-edps@webofconferences.org

Additional copies of this publication are available from:

Curran Associates, Inc.
57 Morehouse Lane
Red Hook, NY 12571 USA
Phone: 845-758-0400
Fax: 845-758-2633
Email: curran@proceedings.com
Web: www.proceedings.com

TABLE OF CONTENTS

Statement of Peer review

In submitting conference proceedings to *Web of Conferences*, the editors of the proceedings certify to the Publisher that

1. They adhere to its **Policy on Publishing Integrity** in order to safeguard good scientific practice in publishing.
2. All articles have been subjected to peer review administered by the proceedings editors.
3. Reviews have been conducted by expert referees, who have been requested to provide unbiased and constructive comments aimed, whenever possible, at improving the work.
4. Proceedings editors have taken all reasonable steps to ensure the quality of the materials they publish and their decision to accept or reject a paper for publication has been based only on the merits of the work and the relevance to the journal.

Title, date and place of the conference

International Conference on Electronics, Engineering Physics and Earth Science (EEPES 2023)

21st-23rd June, 2023
Kavala, Greece

Proceedings editor(s):

Prof. Dr. Teodor Iliev, University of Ruse, Bulgaria
Prof. Dr. Ivaylo Stoyanov, University of Ruse, Bulgaria
Prof. Dr. Grigor Mihaylov, University of Telecommunications and Post, Bulgaria
Dr. Jacob Fantidis, International Hellenic University, Greece

Date and editor's signature **06 July 2023**

Preface

Teodor Iliev[1*], *Ivaylo* Stoyanov[2], *Grigor* Mihaylov[3], and *Jacob* Fantidis[4]

[1]University of Ruse, Department of Telecommunication, 7004 Ruse, Bulgaria
[2]University of Ruse, Department of Electrical Power Engineering, 7004 Ruse, Bulgaria
[3]University of Telecommunications and Post, Department of Telecommunication, 1000 Sofia, Bulgaria
[4]International Hellenic University, Department of Physics, 65404 Kavala, Greece

Abstract. International Conference on Electronics, Engineering Physics and Earth Science (EEPES 2023), 21 - 23 June, 2023, Kavala, Greece.

The second edition of the International Conference on Electronics, Engineering Physics and Earth Science (EEPES 2023) was held in hybrid mode in Kavala, Greece from 21 to 23 June, 2023.

The EEPES 2023 Conference is the premier international venue focused on understanding the current state and the evolution of the science through the lens of engineering physics, electronics devices and microelectronics, telecommunications, optical and electromagnetic communication, artificial Intelligence and soft computing, environmental physics and renewable energy and many other disciplines. It is a place where researchers, academics, people in industry, students and the alike to present and discuss progress in research, development, standards, and applications related to the applied and engineering physics.

The technical program of this year's edition of the conference consists of two distinguished keynotes by world renowned experts with their plenary lectures as follows:

The technical program of this year's edition of the conference consists of two distinguished keynotes by world renowned experts with their plenary lectures as follows:

- Prof. Dr. George A. Papakostas, International Hellenic University, Greece, the title of the lecture "Computer Vision in the Industry Era";
- Prof. Dr. Veneta Aleksieva, Technical University of Varna, Bulgaria, the title of the lecture "Blockchain Based Solutions: Industry's Dreams or Reality".

We have prepared interesting plenary sessions not only from respected universities professors and academia speakers on visionary topics but also from our sponsors Raycap S.A., Greece and EN.PO.KA S.A., Kavala, Greece had done a distinguished technical presentations.

The technical program of the conference consists of one plenary session, one technical session, and seven in person and nine online tracks for research papers. The authors of the accepted papers, who are unable to attend the conference in person, were given the option to present their papers online and attend an interactive virtual conference. Online presentations

* Corresponding author: tiliev@uni-ruse.bg

were held in a virtual room environment via Zoom Meeting platform for the opening ceremony, all conference sessions, and meetings.

Finally, we are thankful to Municipality of Kavala, Greece and Dimofeleia, Kavala, Greece that becoming am official sponsor of the EEPES 2023 and all above mentioned partners for good cooperation and also for the technical and financial support. Thanks to Institutional partners University of Ruse "Angel Kanchev" and University of Telecommunications and Post, Sofia, both from Bulgaria and International Hellenic University, Greece, TPC members, conference chairs, and reviewers for their wonderful efforts in making a successful and rigorous technical program. Without them, this event would have never happened.

Modeling of pollutant emissions from the turbine engine of the main propulsion of a vessel in operating conditions

Pawel Wirkowski[1], *Jaroslaw* Markowski[2*], *Pawel* Imilkowski[2], *Iliya* Iliev[3], *Krzysztof* Jesionek[4], *Janusz* Badur[5], *Piotr* Dlugiewicz[6], *Jacek* Madry[7], *Grzegorz* Wieczorkiewicz[8] and *Pawel* Benedict[2]

[1]Polish Naval Academy of the Heroes of Westerplatte, Faculty of Mechanical and Electrical Engineering, Smidowicza Street 69, 81-127 Gdynia, Poland
[2]Poznan University of Technology, Faculty of Mechanical Engineering, Piotrowo 3, 60-965 Poznan, Poland
[3]University of Ruse, Department of Heat, Hydraulics and Environmental Engineering, Studentska 8, 7004 Ruse, Bulgaria
[4]Vitelon Collegium State University, Sejmowa 5A, 59-220 Legnica, Poland
[5]Institute of Flow Machines of Robert Szewalski Polish Academy of Sciences, Fiszera 14, 80-231 Gdansk, Poland
[6]Imagine Flying LLC, 116 Cypress Lagoon Ct, 32082 Ponte Vedra Beach, FL, USA
[7]SpinTechSystem Sp. z o.o. Dzialynskich 11, 63-000 Sroda Wielkopolska, Poland
[8]Lower Silesian Center of Modern Technologies Sp. z o.o. Walbrzyska 49/3, 52-314 Wroclaw, Poland

Abstract. The intensification of environmental protection measures is being implemented in all economic and industrial activities. The goal of reducing harmful emissions from various types of combustion engines used in different applications is being pursued dynamically. The introduced changes concern the values of harmful emission limits and the methods of measuring or assessing their emissions. There are solutions for propulsion systems, such as turbine engines used on ships, where there is a need for constant monitoring of operational and environmental parameters during the operation of the vessels. Continuous measurement of harmful emission values contained in exhaust gases during the operation of these systems allows for an assessment of their impact on the environment. However, due to the costs of measuring equipment and limitations associated with its use, alternative methods are being sought to evaluate pollution emissions resulting from the operation of turbine engines. One possibility is to assess the emission characteristics of the propulsion system in emission tests under conditions of its actual operation. Based on this, by introducing basic mathematical tools, it is possible to determine the values of emissions of individual pollutants during the propulsion system operation in actual conditions. These actions are presented in this paper.

* Corresponding author: jaroslaw.markowski@put.poznan.pl

1 Introduction

Turbine internal combustion engines (ICE) for marine applications, due to their properties, constitute only a few percent share in the entire population of ICE used in the propulsion of vessels. Therefore, the existing legal regulations regarding the emission of harmful compounds in the exhaust of marine engines apply only to piston engines. One such piece of legislation is the aforementioned MARPOL Convention. In 1997, at a Diplomatic Conference organized by the International Maritime Organisation (IMO), a Protocol amending MARPOL 73/78 was adopted, which added Annex IV containing provisions for the prevention of air pollution from ships. In view of the above-mentioned legal acts concerning the protection of the marine environment, it can be concluded that the evolution of legislation in the last dozen or so years in the field of setting permissible levels of harmful exhaust emissions from marine engines shows how important it is to harmonize regulations in this area. It is sought that they are used by all vessels, regardless of their purpose. Most of the work related to the analysis of the impact of vessels on the environment, as well as development works carried out as part of various projects, concern primarily civil communication units, transport units and reciprocating compression ignition engines, which are used for their propulsion [1-4]. Currently, due to the requirement of using low-sulfur fuels, ship propulsion systems are being designed with hybrid arrangements, equipped with several power generation units using high-speed diesel engines and with additional sets of turbine engines. When designing these propulsion systems, the volume of the engine room and operating costs associated with periodic maintenance are also taken into account, making TEs(TEs) an increasingly attractive option. As a result, their use in ship propulsion systems may increase in the coming years. Therefore, the work presented in this article focuses on finding ways to continuously monitor and assess the impact of TEsused in ships on the environment. To achieve this, work has been carried out to develop a model for evaluating harmful emissions based on operational parameters.

The issue of assessing emissions of pollutants in the exhaust gases of TEs is largely related to the gas flow parameters and the measurement of the concentration of harmful compounds contained in them. TEs are characterized by a large flow rate of the working fluid through the engine during operation. The working fluid is air to which heat generated in the thermodynamic processes during fuel oxidation in the combustion chamber is supplied. The large airflow used to achieve the required engine operating parameters determines the fuel flow rate supplied. Depending on the energy demand generated by the engine, the processes carried out in the combustion chamber also change, contributing to the change in the concentration of harmful compounds in the exhaust gases. To assess emissions of pollutants in the exhaust gases, the mass flow rate of the exiting gases and the concentrations of individual harmful compounds in them are needed [5]. The concentration of harmful compounds in the exhaust gases was measured with a flue gas analyzer of appropriate measurement accuracy. At the same time, a significant problem is the measurement of the mass flow rate of the exiting gases. Another problem in assessing the emissions of pollutants generated by the main TEs during ship operation is the need to measure the concentration of these compounds during operation. Currently, it is not possible to permanently install flue gas analyzers that will continuously measure the concentration of pollutants due to the costs of purchasing and installing the equipment and operational problems arising from measurement accuracy and the need to provide appropriate operating conditions. Therefore, an analysis was carried out to explore alternative methods. The functional relationships between individual ecological and operating parameters were analyzed to achieve this. The obtained relationships made it possible to determine a kind of individual ecological characteristic of the turbine engine. Tests of harmful emissions in exhaust gases of powered craft with turbine engines.

2 Research on the emission of harmful compounds in the exhaust gases of a watercraft powered by turbine engines

2.1 Test methodology

2.1.1 Measurement object and measuring apparatus for measuring harmful compounds in exhaust gases

The work aimed to develop a method for assessing emissions of harmful exhaust compounds from TEs in use as the main propulsion of a vessel. Therefore, it was necessary to carry out operational tests of the selected unit, during which, in addition to recording the values of engine operating parameters, the concentration values of harmful compounds contained in the exhaust gases were continuously recorded. The research object was the FFG-7 frigate-class ship operated in the Polish Navy. ORP "Gen. T. Kosciuszko" (Figure 1). It is powered by two LM 2500 TEs (Figure 2).

Fig. 1. The warship „Gen. T. Kociuszko".

The basic parameters of the drive train, recorded during operational tests at nominal load, reached the following values:

- vessel's speed (displacement 4,000 t, sea state 1, ambient temperature 293 K) – 30 knots;
- propeller shaft rotation speed – 179 rotation/min;
- propeller pitch – 7.16 m;
- torque on the propeller shaft – 1,594,962 N·m;
- propeller shaft power – 29,630 kW;
- fuel consumption – 7,467 kg/h.

The propulsion system's power is approximately 30,000 kW (i.e., 2 x 15,000 kW) at a maximum turbine speed of n_{TN} = 3,600 rpm. The presented COGAG (Combined Gas Turbine and Gas Turbine) propulsion system allows for operation in three basic variants:

- operation with two engines;
- operation with the engine on the left side;
- operation with the engine on the right side.

The first variant allows one to achieve a speed of about 30 knots, while the other variants allow one to achieve a speed of about 25 knots with only one of the engines. For economic reasons, the second or third operation variant is most often used under normal operating conditions.

LM 2500 engines are turbine ICE with a wide range of applications on modern warships and civilian vessels. The basic structural elements of the engine are:

- 16-stage axial compressor with a maximum pressure of 18:1; the first six stages are equipped with adjustable steering paddles,
- annular combustion chamber equipped with 32 injectors,
- 2-stage exhaust generator turbine with cooled vanes, which is the source of compressor drive and auxiliary mechanisms coupled to the engine,
- 6-stage low-pressure turbine, which is structurally the engine drive turbine.

An integral part of marine structures is an intermediate shaft transmitting the torque of the drive turbine to the reduction gear.

Fig. 2. LM 2500 turbine engine for vessels [6].

Measurements of the energy parameters of the working medium, made in the characteristic control cross-sections of the engine, are an important source of diagnostic information about the state of the structural structure of its flow part. The presented schematic diagram of the LM 2500 engine, with marked control cross-sections of the flow part, makes it possible to illustrate the location of measuring points. The measured and determined engine operating parameters and their measuring range, expressed in units applicable in the ship's engine room, are summarized in Table 1. The concentration of exhaust gas components was measured with the Semtech-DS exhaust gas analyzer (Figure 3) from Sensors (Sensors EMission TECHnology). The characteristics of the Semtech-DS analyzer are presented below (Table 2).

The analyzer allows for the measurement of the concentration of harmful compounds such as carbon monoxide, hydrocarbons, nitrogen oxides, and carbon dioxide. The main advantage of the analyzer is its compact design and small size, which allows for easy movement and positioning in objects being tested under real operating conditions. The analyzer meets the requirements of standard 1065 for measuring exhaust emissions using PEMS systems. It consists of measurement modules:

- FID (Flame Ionization Detector), used to determine the total concentration of hydrocarbons referred to as HC or THC (Total Hydrocarbons) in the exhaust gas,
- NDUV (Non-Dispersive Ultraviolet) type analyzer, non-dispersible to ultraviolet radiation for measuring the concentration of nitric oxide and nitrogen dioxide,

- NDIR (Non-Dispersive Infrared) type analyzer, non-dispersible to infrared radiation for measuring the concentration of carbon monoxide and carbon dioxide,
- ECS (Electrochemical Sensor) analyzer for determining the oxygen concentration in the exhaust gas.

Table 1. LM 2500 engine performance.

Parameter name	Symbol	Unit	Measurement range
Ambient pressure	p_o	[hPa]	800 ÷ 1,040
Ambient temperature	t_o	[°C]	–40 ÷ 40
Rotational speed of the exhaust generator	n_{GG}	[obr/min]	0 ÷ 12,000
Rotational speed of the drive turbine	n_{PT}	[obr/min]	0 ÷ 5,000
Engine inlet air temperature	t_1	[°F]	–40 ÷ 150 (–40 ÷ 65.6°C)
Total air pressure at engine inlet	p^*_1	[psig]	0 ÷ 16 (0 ÷ 1.103 bar)
Air pressure downstream of the compressor	p_2	[psig]	0 ÷ 300 (0 ÷ 20.684 bar)
Exhaust gas temperature in front of the drive turbine	$t_{4.2}$	[°F]	0 ÷ 2,000 (–18 ÷ 1,093°C)
Total exhaust gas pressure upstream of drive turbine	$p^*_{4.2}$	[psig]	0 ÷ 75 (0 ÷ 5.171 bar)
Exhaust gas temperature	t_6	[°F]	0 ÷ 1,000 (–18 ÷ 538°C)
Fuel temperature in front of the engine	t_{fuel}	[°F]	0 ÷ 100 (–18 ÷ 37.8°C)
Fuel pressure before injectors	p_{fuel}	[psig]	0 ÷ 1,500 (0 ÷ 103.421 bar)
Torque on the drive turbine shaft	M_{PT}	[LB·FT]	0 ÷ 50,000 (0 ÷ 67.770 kN·m)
Power on the drive turbine shaft	P_{TN}	[KM]	0 ÷ 25,000 (0 ÷ 18,642 kW)

Fig. 3. Exhaust gas analyzer.

The exhaust gases are introduced into the analyzer using a measurement probe that maintains a temperature of 191°C. In the next step, solid particles are filtered out, and the concentration of hydrocarbons is measured using a flame ionization analyzer. Then, the exhaust gases are cooled to a temperature of 4°C, and the concentration of nitrogen oxides is measured using a non-dispersive method using ultraviolet radiation that allows for simultaneous measurement of nitrogen oxide and nitrogen dioxide concentrations.

Table 2. Characteristics of the Semtech-DS analyzer [101].

Parameter	Method of measurement	Precision
Concentration of compounds in exhaust gases: CO HC NOx = (NO+NO2) CO2 O2	 NDIR, measuring range 0 ÷ 10% FID, range 0 ÷ 10,000 ppm NDUV range 0 ÷ 3,000 ppm NDIR range 0 ÷ 20% Electrochemical analyzer, range 0 ÷ 20%	 ±3% of the measuring range ±2.5% of measuring range ±3% of the measuring range ±3% of the measuring range ±1% of measuring range
Exhaust flow	Mass flow rate flue gas temperature up to 700 oC	±2.5% of measuring range ±1% of measuring range
Warm-up time	900 s	-
Response time	T90 < 1 s	-

The exhaust gases are introduced into the analyzer using a measurement probe that maintains a temperature of 191°C. In the next step, solid particles are filtered out, and the concentration of hydrocarbons is measured using a flame ionization analyzer. Then, the exhaust gases are cooled to a temperature of 4°C, and the concentration of nitrogen oxides is measured using a non-dispersive method using ultraviolet radiation that allows for simultaneous measurement of nitrogen oxide and nitrogen dioxide concentrations. Carbon monoxide, carbon dioxide, and oxygen concentrations are measured using a non-dispersive method using infrared radiation and an electrochemical analyzer, respectively. In addition to measuring exhaust gas components, this analyzer also allows for measuring the mass flow rate of the exhaust gases. Data can be directly transmitted to the analyzer's central unit from the diagnostic system of the tested object, and GPS location signals can also be used.

2.1.2 Method of measurement of harmful compounds in exhaust gases

The research was conducted on a vessel powered by LM 2500 engines under operational conditions. The flue gas analyzer probe was placed in the axis of the final outlet of the exhaust system in the upper part of the ship superstructure (Figure 4). Additionally, the location of the sampling point of the exhaust gas was about 15 meters away from the turbine outlet, ensuring sufficient mixing of the exhaust gas, and therefore the sampling of a representative sample.

Fig. 4. Superstructure on the upper deck with the exhaust system and view of the probe mounted in the exhaust outlet.

The measurements were carried out for the engine load range from idle speed to the maximum load achievable during the tests, with the rotational speed of the exhaust gas generator (n_{GG}) set as the parameter determining the measurement points. This parameter was obtained by appropriate changes in the fuel flow rate supplied to the engine (m_{pal}), which caused the change in the rotational speed of the power turbine (n_{PT}), and the entire propeller shaft (n_{LW}), and finally had an impact on the pitch of the propeller blades. In this way, the resulting ship's speed was achieved. The tests were conducted in such a way that the desired values of n_{GG} were set first. After obtaining the specified engine operating parameters, the selected operating parameters of the propulsion system were recorded and saved. The engine operated at a set load for 300 seconds. The measurement time for both the operating parameters of the propulsion system and the exhaust gas parameters was 20 seconds with a sampling frequency of 1 Hz. The obtained averaged values of the measured operating parameters were used to conduct the necessary analyses to determine the emission characteristics of the turbine engines.

2.2 Emission characteristics

Based on the obtained results, analyses were conducted, resulting in emission characteristics of the TEs of the propulsion system. To make them more efficient, they are presented as a function of the load indicator (Figure 5 and Figure 6).

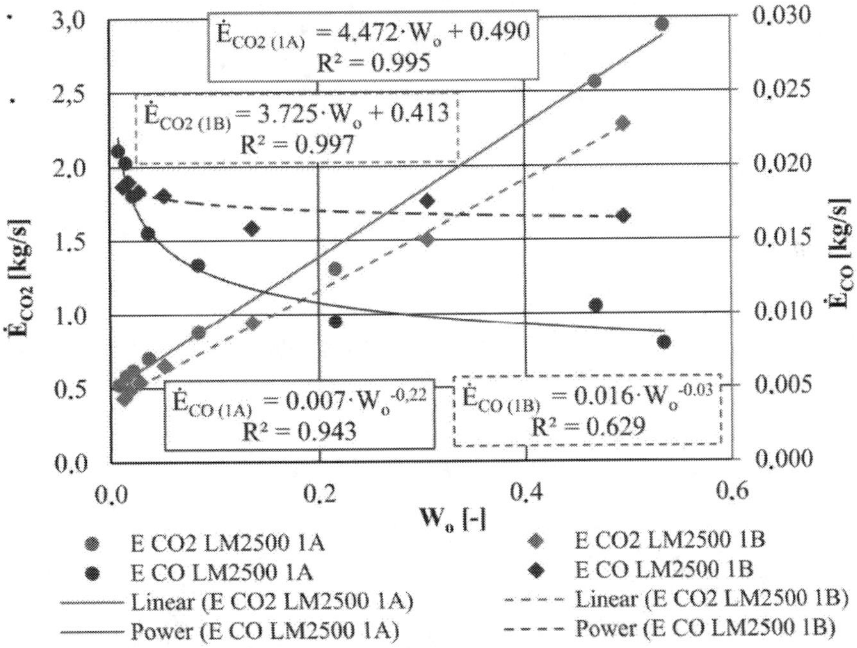

Fig. 5. Carbon dioxide (E_CO2) and carbon monoxide (E_CO) emissions mass load of LM 2500 1A and 1B engines as a function of engine load factor (W_o)

Having the values of the mass emission intensity of the analyzed harmful exhaust compounds of the LM 2500 1A and LM 2500 1B engines and knowing the duration of each measurement, which was 300 seconds, it is possible to calculate the value of the mass of the emitted harmful compounds during each measurement, as well as for the entire measurement test of the engine. The value of the mass of the emitted harmful compound for a given measurement (engine load factor) was determined from the relationship:

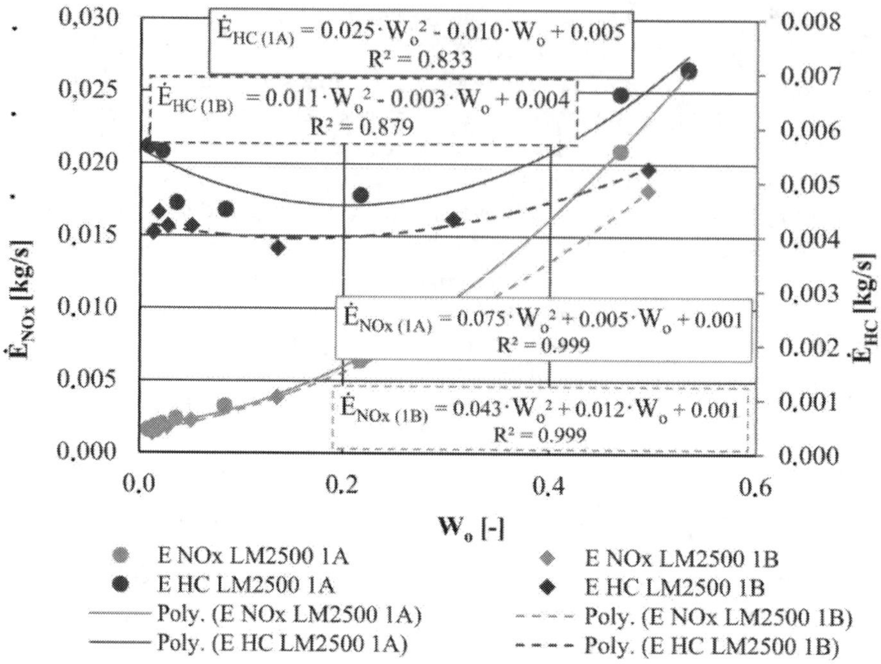

Fig. 6. The mass emission rate of nitrogen oxides (E˙_NOx) and hydrocarbons (E˙_HC) for LM 2500 1A and 1B gas turbines as a function of engine load factor (Wo).

$$m_{i,j}(W_o) = \dot{E}_{i,j}(W_o) \cdot \tau_j [kg], \tag{1}$$

where:

$\dot{E}_{i,j}(W_o)$ – mass emission rate of the i-th component of the exhaust gas [kg/s] as a function of the engine load factor for the j-th measurement,

τ_j – duration of the j-th measurement [s]

The total mass of the pollutant emitted during the engine measurement test was determined as the sum of the masses of the compound emitted for each measurement:

$$m_i = \sum m_{i,j}(W_o) \, [kg]. \tag{2}$$

The values of the mass of the emitted harmful compounds can thus be determined continuously during operation or in summary terms from the entire service period, e.g., a cruise.

Therefore, the presented methodology for assessing the ecological indicators of the operation of a warship powered by TEs can be improved by using a numerical algorithm, which can be applied in various programming languages to obtain application tools. For example, an algorithm was prepared to demonstrate the developed emission model.

3 Model of pollutant emissions from the turbine engine of the main propulsion of a vessel in operating conditions

The next step after performing the research on the real object was to create a mathematical model that is its theoretical reflection. This model will allow testing the validity of the assumptions used in testing the object, and, moreover, will allow to check the reliability of the results obtained at a later time. This task was carried out using the SciLab software, and the model is presented in Figure 7 and Figure 8. Thanks to the previously developed emission characteristics of the LM2500 1A engine, in order to obtain information of interest to us from the generated mathematical model, the only data necessary to enter is the power generated by the engine in each second of the considered movement. The measurement results were saved in a file in the .xls format.

The developed mathematical model starts its operation by downloading data from the previously mentioned file, in which information on the power generated by the LM2500 1A motor with a rated power of 24608 kW has been archived. These data are saved in the form of a matrix, which will be later used to calculate the engine load factor (W_0 in SciLab), the momentary emission of individual compounds (M_CO, M_CO2, M_NOX, M_HC in SciLab) and the total mass of emitted harmful compounds (Calk_M_CO , Calk_M_CO2, Calk_M_NOX, Calk_M_HC). At this point, it should be noted that in order to carry out the above-mentioned calculations, it was necessary to use the unit emissions of individual compounds obtained as a result of measurements and to develop emission trend lines for each of them. Calculations of specific emissions depending on the engine load factor are performed in the program in lines from 31 to 34. The results of the calculations are presented in the Figure 9, Figure 10 and Figure 11.

```
1  clear()
2  clc()
3  //Decode file, extract and open Excel stream
4  [fd,SST,Sheetnames,Sheetpos] = xls_open('Motion_record_EN.xls');
5  //Read data sheet
6  [Value,TextInd] = xls_read(fd,Sheetpos(1));
7  //Close the spreadsheet stream
8  mclose(fd);
9
10 // INPUT DATA LM2500 ENGINE
11 //RATED ENGINE POWER [kW]
12 P_max=24608
13
14 //Measurement time [s] - each measurement is performed every 1 second
15 t_pomiaru=1
16
17 //Saving data from the measurement protocol (FROM EXCEL)
18 P_act=Value(2:308,2);
19 t_calk=Value(2:308,1);
20
21 // Determining the size of the matrix/number of elements for the "for loop"
22 number_of_elements= size(P_act,1)
23
24 for x=1:number_of_elements
25
26     //Calculation of engine load value
27     W_0(x,1)=P_act(x,1)/P_max;
28
29     //Emission of harmful compounds [kg/kWh] (characteristics taken
30     //from measurements carried on a real object)
31     e_CO = 0.0011*(W_0^(-1.225));
32     e_CO2 = 0.4498*(W_0^(-0.587));
33     e_NOX = 0.0039*(W_0^(-0.331));
34     e_HC = 0.0009*(W_0^(-0.966));
35
36     //Measurement of momentary emission of harmful compounds
37     M_CO(x,1)=(W_0(x,1)*P_max*t_pomiaru*e_CO(x,1))/3600
38     M_CO2(x,1)=(W_0(x,1)*P_max*t_pomiaru*e_CO2(x,1))/3600
39     M_NOX(x,1)=(W_0(x,1)*P_max*t_pomiaru*e_NOX(x,1))/3600
40     M_HC(x,1)=(W_0(x,1)*P_max*t_pomiaru*e_HC(x,1))/3600
41
```

Fig.7. The program, written in the SciLab environment, is designed to analyze the level of emission of harmful compounds during the examination of the object.

 https://doi.org/10.1051/e3sconf/202340401001

```
    //Measurement of the total emission of
    //harmful compounds during the measurement
    if x==1 then,
        Calk_M_CO(x,1)= M_CO(x,1);
        Calk_M_CO2(x,1)= M_CO2(x,1);
        Calk_M_NOX(x,1)= M_NOX(x,1);
        Calk_M_HC(x,1)= M_HC(x,1);
    else
        Calk_M_CO(x,1)= Calk_M_CO(x-1,1) + M_CO(x,1);
        Calk_M_CO2(x,1)= Calk_M_CO2(x-1,1) + M_CO2(x,1);
        Calk_M_NOX(x,1)= Calk_M_NOX(x-1,1) + M_NOX(x,1);
        Calk_M_HC(x,1)= Calk_M_HC(x-1,1) + M_HC(x,1);
    end
end

//Charts
//Graph of momentary emission of harmful compounds
gda().grid = [1 1]*color("grey70");
title(gda(), "fontsize", 3, "color", "lightseagreen", "fontname", "helvetica bold");
scf(1); clf


subplot(2,2,1), plot(t_calk, M_CO)
title("Momentary CO emission as a function of time");
xlabel "Time [s]"
ylabel "CO [kg/s]"

subplot(2,2,2), plot(t_calk, M_CO2)
title("Momentary CO2 emission as a function of time"); gca().sub_ticks(1) = 8;
xlabel "Time [s]"
ylabel "CO2 [kg/s]"

subplot(2,2,3), plot(t_calk, M_NOX)
title("Momentary NOx emission as a function of time"); gca().sub_ticks(2) = 8;
xlabel "Time [s]"
ylabel "NOx [kg/s]"

subplot(2,2,4), plot(t_calk, M_HC)
title("Momentary HC emission as a function of time"); gca().sub_ticks = [8 8];
xlabel "Time [s]"
ylabel "HC [kg/s]"
sda();

//Graph of the total emission of harmful compounds during the measurement
gda().grid = [1 1]*color("grey70");
title(gda(), "fontsize", 3, "color", "lightseagreen", "fontname", "helvetica bold");
scf(2); clf

subplot(2,2,1), plot(t_calk, Calk_M_CO),
title("Total CO emissions at the time of measurement");
xlabel "Time [s]"
ylabel "CO [kg]"

subplot(2,2,2), plot(t_calk, Calk_M_CO2),
title("Total CO2 emissions at the time of measurement"); gca().sub_ticks(1) = 8;
xlabel "Time [s]"
ylabel "CO2 [kg]"

subplot(2,2,3), plot(t_calk, Calk_M_NOX),
title("Total NOx emissions at the time of measurement"); gca().sub_ticks(2) = 8;
xlabel "Time [s]"
ylabel "NOx [kg]"

subplot(2,2,4), plot(t_calk, Calk_M_HC),
title("Total HC emissions at the time of measurement"); gca().sub_ticks = [8 8];
xlabel "Time [s]"
ylabel "HC [kg]"

//Graph of the emission of harmful compounds in the aspect of engine load value
gda().grid = [1 1]*color("grey70");
title(gda(), "fontsize", 3, "color", "lightseagreen", "fontname", "helvetica bold");
scf(3); clf

subplot(2,2,1), plot(W_0, M_CO),
title("Total CO emissions at the time of measurement");
xlabel "Engine load [-]"
ylabel "CO [kg/s]"

subplot(2,2,2), plot(W_0, M_CO2),
title("Total CO2 emissions at the time of measurement"); gca().sub_ticks(1) = 8;
```

Fig.8. Further part of the program, written in the SciLab environment, is designed to analyze the level of emission of harmful compounds during the examination of the object.

Figure 9 shows the momentary emissions of individual compounds emitted during the operation of the LM2500 1A combustion engine. The engine in the simulation presented in Figure 9, Figure 10 and Figure 11 worked as follows - from the level of 1,000 kW of generated power, it increased it by 100 kW every second until reaching the maximum power of 24,608 kW. This power was maintained for 13 seconds, after which it began to drop by 500 kW every second. The whole considered movement lasted 307 seconds, the increase of power took place for 238 seconds, and after 13 seconds of stability, it decreased for the next 56 seconds. The presented graphs show that with the increase of the power generated by the engine, and thus the load, the CO_2, NO_x and HC emissions increase, while the CO emission decreases significantly.

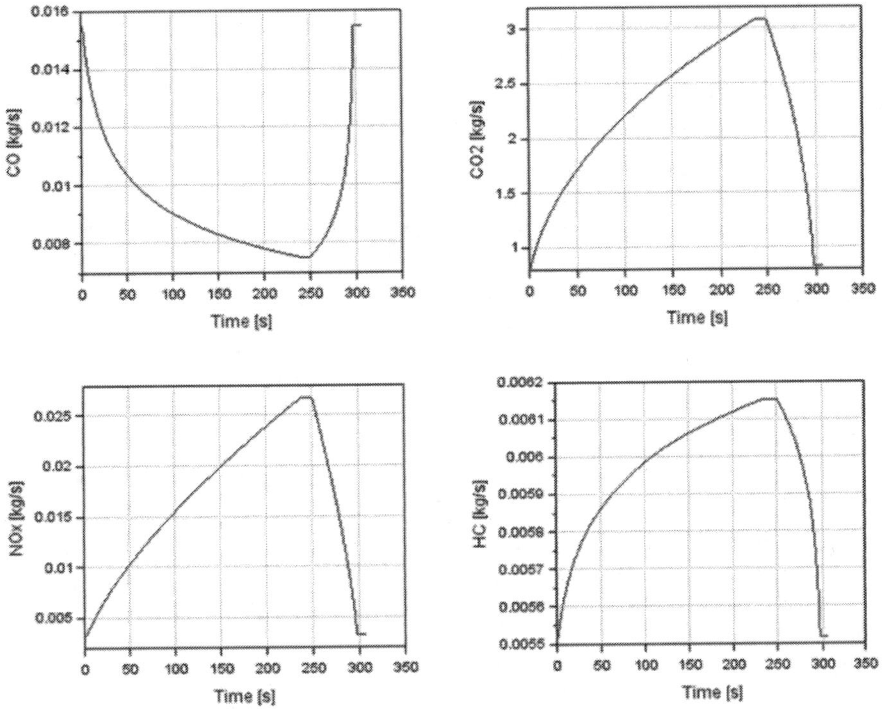

Fig. 9. Momentary emission of harmful compounds depending on the duration of operation of the considered object.

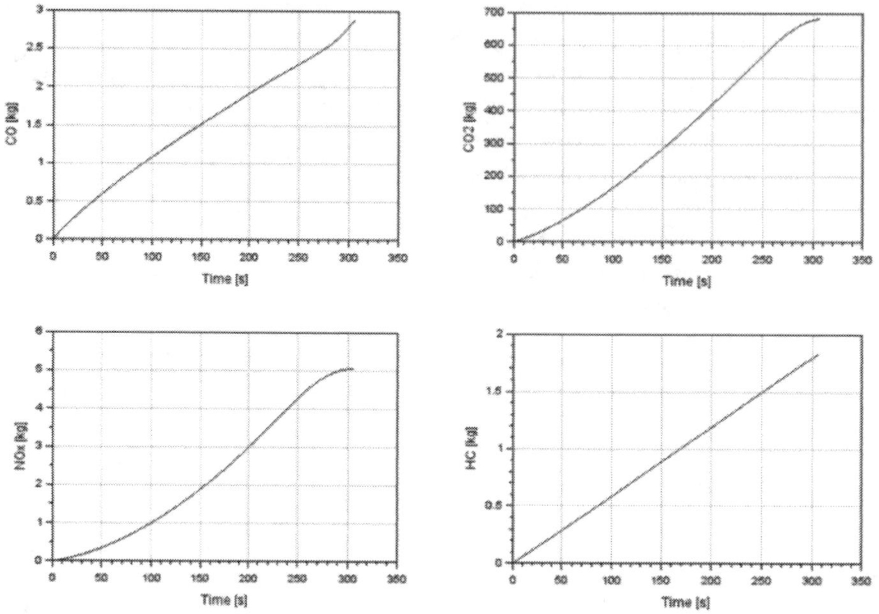

Fig. 10. Total emission of harmful compounds depending on the duration of operation of the considered object.

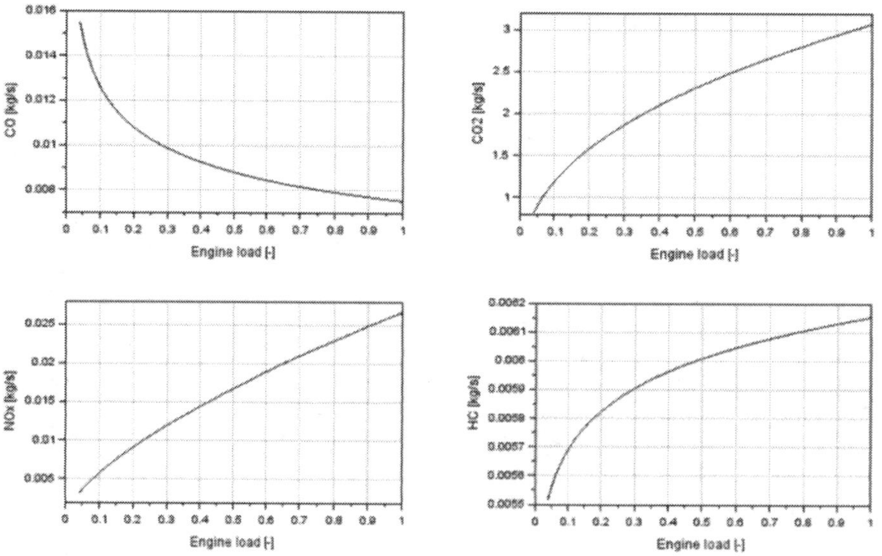

Fig. 11. Momentary emission of harmful compounds depending on the engine load factor of the considered object.

4 Conclusions

The developed method of assessing the ecological parameters of the propulsion system of warships is conditioned by the characteristics of the emission parameters of such vessels. These characteristics must be determined first, as they, together with all components of the propulsion systems, constitute the ecological assessment of the warship as an exploited object. Continuous measurement of several basic operating parameters, along with their relation to the emission characteristics of the particular vessel, allows for a quick determination of the current emissions of pollutants. Additionally, the use of algorithms implemented in numerical tools enables a quick assessment of the impact of the warship on the environment. The proposed tool allows for a continuous evaluation of emissions, which, combined with constant monitoring of operating parameters, can be used to optimize the usage conditions of warships and consequently contribute to reducing CO_2 emissions into the atmosphere. However, it should be mentioned, that the numerical model was developed on the basis of designated trend lines, which describe the phenomenon to which they refer with some approximation. The measure of the credibility of the trend line is the R^2 coefficient. The closer its value is to 1, the better the trend line is suited to the described phenomenon. For this reason, the mathematical model is only an approximation of the real object.

This work has been accomplished with financial support by Grant No BG05M2OP001-1.002-0011-C02 financed by the Science and Education for Smart Growth Operational Program (2014-2020) and co-financed by the European Union through the European Structural and Investment funds.

References

1. T. Hasegawa, *Progress in Gas Turbine Performance, Development of Semiclosed Cycle Gas Turbine for Oxy-Fuel IGCC Power Generation with CO2 Capture*, InTech open science open minds, (2014)
2. J. Herdzik, *Emissions from marine engines versus IMO certification and requirements of Tier 3.* Journal of KONES Powertrain and Transport, Vol. **18**, 2 (2011)
3. J. Markowski, J. Pielecha, R. Jasiński, P. Wirkowski, G. Ślusarz, *Evaluation of Relations Operating and Ecological Parameters of Turbine Engines*, Journal of Polish CIMEEAC, 10 (2015)
4. J. Markowski, D. Olejniczak, P. Wirkowski, *Evaluation of turbine microjet engine operating parameters in conditions conducive to inlet freezing*, MATEC Web of Conferences, **118** (2017)
5. P. Wirkowski, J. Markowski, T. Kniaziewicz, *Przygotowanie do badań emisji związków szkodliwych w spalinach okrętowego turbinowego silnika w warunkach eksploatacji*, Journal of Polish CIMEEAC, Vol. 12, 1 (2017)
6. https://www.akaero.com.tr/?l=tr&p=urun_bilgi&id=12 (Accesed on 15 April, 2023)

A concept for flood early warning

Iskra Simova[*], *Borislav Ganev*, *Rositsa Velichkova* and *Elitsa Gieva*

Technical University of Sofia, 1000 Sofia, Bulgaria

Abstract. Bulgaria is exposed to a number of natural hazards and risks negative consequences have a significant impact on the environment and the population, industry, infrastructure, cultural heritage, etc. Floods are the second most common disaster in Bulgaria. In the present work a statistical analysis of the floods in Bulgaria for the period 2010-2020 is made. The present paper also proposes an idea for an early warning system for floods. The idea is to work on early forecasting, which will help to timely notify municipalities and regional fire departments in order to quickly control the floods and its consequences.

1 Introduction

A flood is a natural hazard, a natural phenomenon that is the result of heavy rainfall, intense snowmelt, ice, sliding stones and lumps causing flooding of riverbeds, rains that uproot trees, destroy buildings, etc. They can also be caused by huge waves (tsunamis) along the ocean coast due to earthquakes, volcanic activity, cyclones, tornadoes, storms, etc. Major flooding of our inland rivers dates back centuries due to torrential rains.

Although flooding is a natural phenomenon, human activity and intervention can be significant, such as alteration of natural riverbeds related to urbanization, agricultural activities and deforestation, changes in river basin conditions, etc. Earth's climate is changing rapidly and the likelihood of flooding is expected to increase. In recent years, floods in Europe caused a lot of damage - loss of human life, property, crops, etc. A flood is a natural disaster in which parts of the earth's surface are inundated with water [1]. Unlike other natural disasters, floods are highly predictable both in terms of their occurrence and distribution, and in terms of possible consequences. The floods appearance is largely influenced by human activity and human intervention in natural processes. During the last decade, the problems related to flood appearance in Bulgaria have been intensively studied [2, 3, 4]. The aim of the current paper is to achieve a positive impact, sustainable change and the prevention and assessment of flood risk, management and mitigation of the consequences of floods in the territory of Bulgaria.

2 Floods in Bulgaria

European Commission's guidelines contained in the "Draft list of flood types and stand consequence" from 16.02.2011 distinguish some types of floods according to their source:

* Corresponding author: isimova@tu-sofia.bg

- River floods – inundation of land by water from natural or improved drainage systems. Such floods are caused by rivers, drains, mountain streams, short-term one-off and/or periodic flows (e.g. inland), etc.
- Flooding from heavy rainfall – due to heavy rainfall, the drainage infrastructure in densely populated areas or the natural ability of soil outside populated areas to absorb rainwater cannot absorb the large volumes of water.
- Groundwater flooding – groundwater rushes to the surface. This process may be due to a sudden increase in groundwater, which is also often associated with high surface water levels.
- Marine flooding – Flooding from seawater, estuaries or coastal lakes. This flooding can be caused by extreme tidal levels, sustained high winds, and large waves such as coastal tsunamis.
- Infrastructure flooding – from artificial reservoirs or drainage facilities. They can be caused by accidents or insufficient capacity of reservoirs (dams), treatment plants, water supply and sewage facilities.
- Snowmelt flooding - can be combined with precipitation or ice jamming/blocking.

According to the mechanism of occurrence, floods are divided into:

- Natural overflow - inflow of water above the water level capacity of a drainage, river bed or bank.
- Spill protection structures – e.g. spillways or embankments.
- Failure of protective or infrastructural facilities – for example, breaching and failure of a dam wall or failure of pumping equipment.
- Blockage/Retention of water – due to blockage of the natural path of water drainage such as silting under bridges from natural sediments or municipal waste. According to the frequency of occurrence, floods could be divided into:
- Flash flood – occurs quickly and without the possibility of prediction. It is usually due to intense rainfall over a relatively small area.
- A flood with an average rate of occurrence – occurs more slowly than a flash flood and has some chance of being predicted.
- Slow-onset flooding – takes longer than average to develop. It can be predicted relatively early.

Depending on the intensity, recurrence and impact on the environment, hydrometeorologists give the following classification of floods [1]

1. A small flood occurs once every 10-20 years and the lowest part of the riverbed flows out without causing damage.
2. Dangerous floods – every 20-40 years the river submerges the adjacent river terrain in a large section. They can cause damage and pose a danger to people and animals in the immediate vicinity.
3. Highly dangerous floods - They happen every 40-80 years. Rivers flood a lot at dangerous depths (over 1 m) and velocities (over 20 m/s). Damage to bridges, culverts, retaining walls, adjacent land, buildings, etc. They pose a major threat to coastal residents.
4. Extraordinary floods - every 80-150 years the river submerges all or most of the river valley, the water depth exceeds 1.5-2 meters, the flow speed is 2.5-3 meters per second and the river instead flows along the valley of the river bed and bridges that undergo major changes, dams, buildings, hydrotechnical facilities are destroyed. They cause enormous material damage and casualties.
5. Catastrophic floods - 1 time in 150-200 years. Huge territories are flooded, the river valley is submerged and the water flow is completely oriented along it.

https://doi.org/10.1051/e3sconf/202340401002

3 Assessment of floods in period 2010-2020

According to the Water Act, in Bulgaria there are defined spate areas called "Project units". The methodology sets criteria for determining the boundaries of the project units with aim to optimize the work. Bases for determining the project units are major river basins, some of which are divided into two or more units. The main criteria for separation of project units are the areas of the basins (Figure 1).

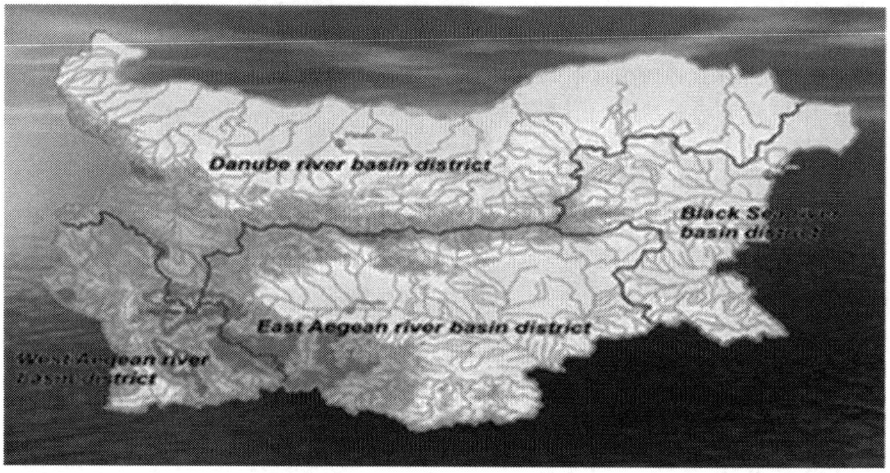

Fig. 1. Map of the Basin Districts in Bulgaria.

Each of this basin district has a number of river. The name of the river and their catchment is given at Table 1.

Table 1 Rivers and river catchment in each district area [5].

Region	River	River catchment km2
West Aegean	Struma	8545
	Mesta	2789
	Dospat	636
Danube	Erma	436
	Nishava	701
	river west of Ogosta	3910
	Ogosta	4282
	Iskar	8633
	Vit	3227
	Osam	2838
	Yantra	7861
	Rusenski Lom	2985
	Danube Dobrudzha river	8140
	Danube	4217
East Aegean	Maritsa	34166
	Tundzha	8429
	Arda	5795
Black Sea	Batova	339
	Provadiya	2131
	Kamchiya	5358
	Severnoburgas	2185
	Mandren	985
	Yuzhnoburgaski	133
	Veleka	995

 https://doi.org/10.1051/e3sconf/202340401002

	Rezova	183

Distribution of the number of flood in period 2010-2020 per basin district is given at Figure 2 [5].

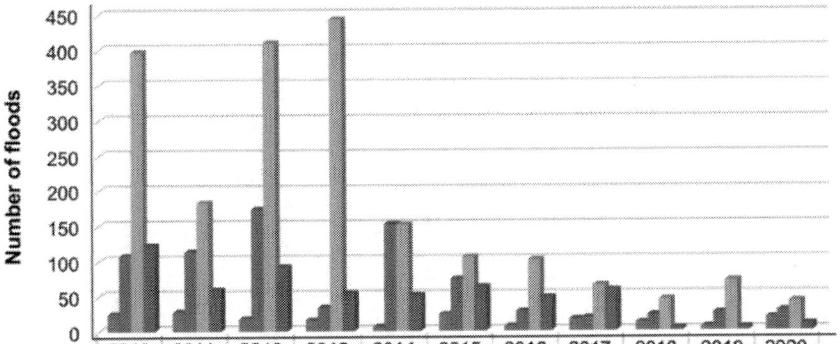

	2010	2011	2012	2013	2014	2015	2016	2017	2018	2019	2020
West Aegean river	25	28	18	16	6	24	7	17	12	6	19
Danube river	107	113	173	34	152	74	28	19	23	26	29
East Aegean river	397	182	410	443	151	105	101	65	45	71	42
Black Sea river	122	59	91	54	51	63	48	58	4	5	10

Year

West Aegean river ■ Danube river ■ East Aegean river ■ Black Sea river

Fig. 2. Floods in Bulgaria in the period 2010-2020 [5].

The most of the floods are in East Aegean river region followed by Danube river region. This can be explained by the fact that we have rivers with the largest river catchment, which is also observed in the second case. There are least floods in West Aegean rive region where the smallest watershed is located. There has also been a decrease in floods in recent years, which can be attributed to the fact of better awareness among the population. In the next section, an integrated system for early prediction of floods in populated areas is proposed.

In Figure 3 is given the damages which the floods costs. It is very clear that the flood damage does not correspond to the number of floods by region. This shows that depending on the type of flood and its strength, it logically leads to greater losses in BGN. Therefore, one forecasting of the floods would lead to a reduction of the damages from the floods on the territory of the Republic of Bulgaria.

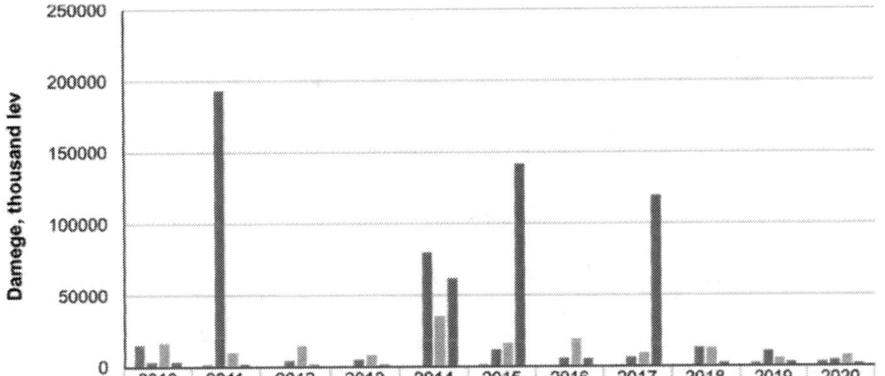

	2010	2011	2012	2013	2014	2015	2016	2017	2018	2019	2020
West Aegean river	15235	1518	515	433	648	1401	88	679	363	2252	3159
Danube river	3496	193346	4248	5169	79914	11675	5788	6286	12891	10508	4324
East Aegean river	16638	9841	14524	8085	35186	16328	19236	9137	12701	5434	7455
Black Sea river	3513	1954	1611	1598	61856	141628	5505	119428	2429	2979	1726

Year

Fig. 3. Damage from floods in period 2010-2020 [5].

4 System for early warning

Based on the data shown for floods and damage in Bulgaria, it can be seen that this is a huge problem for the country. There is a great need for early warning of the population and the services for floods. If the flood is detected earlier the smaller the consequences will be.

At the Figure 4, an early warning system is proposed, which will be placed on bridges before settlements in order to warn the population of flooding. [6-8]

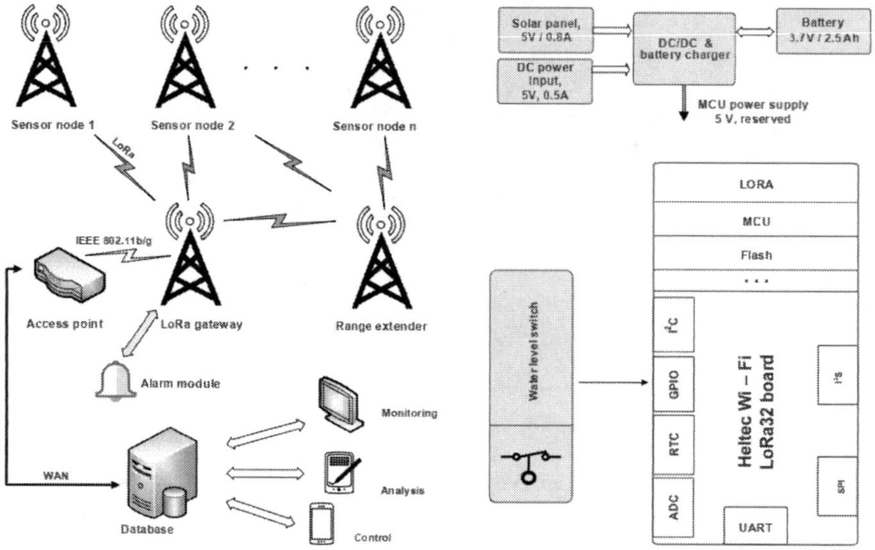

Fig. 4. Scheme of proposed early warning system.

The system is implemented based on spread spectrum transceivers (LoRa), which provides large area coverage while keeping power consumption low. A large number of sensor nodes can be easily added, and one or more range extenders can be added if more distance is needed between them. With the devices we tested, a distance between the sensor node and the gateway of the order of about 10 km is achieved outside of populated areas in direct line-of-sight and about 2.5-3 km in populated areas and in the presence of more buildings. The gateway, besides performing the standard functions of information exchange between the two different network topologies(Wi-Fi and LoRa), in this particular application also triggers the alarm output, which ensures reliable operation of the system even in case of problems or absence of internet connection. The connection to the internet and the recording of the data in a database is more to facilitate remote diagnostics and to ensure system availability. Each of the sensor nodes transmits to the database its unique identification number, the battery charge level and the status of the contact that opens when the water level is high, at a specific programmable time interval (in this case an interval of 30 sec. ÷ 2.5 min.). The identification numbers of all sensor nodes are available at the gateway, so that if there is no communication for more than 5 minutes with any sensor node, a warning is displayed that there is most likely a problem with that particular sensor node, so that adequate measures can be taken to restore the full functionality of the system. When a contact is opened, the sensor node starts transmitting a signal to this effect every 2 seconds, in which case no battery level is monitored and sending the signal has the highest priority. Particular attention has been paid to redundancy of the sensor node power supply. Each node is powered by a high capacity Li-Po battery (2,5Ah) capable of providing autonomous operation of the sensor node for at least 3 days. The battery is charged by a solar panel which is capable of charging it from 10% to 80% in about 4-5 hours. Each sensor node has been developed based on Heltec Wi - Fi LoRa32 board. Range extenders have been implemented with the same board, and here also solar panel power has been used so that they can be installed in remote areas where mains power is not available. The Gateway is installed in a location with internet connectivity and where the decision center is located. Network power is usually available there, so only a battery is needed to ensure power redundancy.

5 Conclusion

An analysis of floods in Bulgaria for period 2010-2020 is given. This can be very helpful for assessment of the risk of floods in Bulgaria. It is presented an idea for early warning systems for floods, which are extremely necessary for our country, and in order to protect people and environment. The proposed integrated warning system can help reduce damage to floods, as well as to warn the population located near places of floods. Incorporating flood risk assessments into planning and in the management of densely populated areas subject to high risk of floods and implementation of damage avoidance measures will contributed to increasing the sustainability of the built environment and others infrastructure.

This research is supported by the Bulgarian Ministry of Education and Science under the National Program "Young Scientists and Postdoctoral Students-2"

References

1. D. Slavov, Pr.Int.Sc.Con.D.T., (2021)
2. V. Spiridonov, S. Balabanova, Climate and Land Use Impacts on Natural and Artificial Systems, (2021)
3. R. Velichkova, IOP Conf. S. M.S.E **878** (2020)
4. R. Velichkova, R.Angelova, I. Simova, E3S W. of C. **207**, (2020)
5. www.nsi.bg
6. S. Antonov, CreBUS 2019; Category number CFP19U17-ART; Code 152084,(2019)
7. D. Stoitseva-Delicheva, V. Karlova-Sergieva, A. Marinchev, V. Galabov, Ch. in H. E. &R, pp. 18, (2019)
8. I. Kralov, J. Genov, I. Angelov, IOP M. S. and En., 878(1), 012036, (2020)

Extensive use of net-metering in Crete's region public buildings

Marini Markaki[1*], *Emmanuel* Karapidakis[1], *Yiannis* Katsigiannis[1], *Maria* Apostolaki[2], *Giannis* Kornilakis[1], and *Panagiotis* Floyris[1]

[1]HMU, Electrical and Computer Engineering Department, 71410 Crete, Greece
[2]Region of Crete, Environment & Spatial Planning Directorate, Crete, Greece

Abstract. Net metering and zero feed-in are two different methods used in photovoltaic (PV) installations to manage the excess energy locally produced by consumers. Net metering allows the energy generated by the PV system to be fed back into the grid, and the excess energy is credited to the owner's account. Zero feed-in, on the other hand, involves the use of an inverter that reduces the amount of energy exported to the grid to zero, while still allowing the PV system to generate power for self-sufficiency use. This paper examines the advantages and disadvantages of both net metering and zero feed-in systems, whereas proposes an algorithm for optimal dimensioning of PV considering the relevant load profiles of each examined load demand. The analysis considers factors such as self-sufficiency and rejection of PV production. The results of this study can help stakeholders to make informed decisions when they are obliged to use a zero feed-in scheme for their PV installations.

1 Introduction

Buildings are responsible for about a third of global energy consumption and a quarter of CO_2 emissions [1]. In particular, the building sector is responsible for approximately 40% of primary energy consumption and 36% of CO_2 emissions in the European Union (EU) [2], where about 35% of the buildings are over 50 years old and 75% of the overall stock is energy inefficient [3]. Consequently, the required energy demand, either in the form of thermal energy (mainly oil) or in the form of electricity, results in significant emissions, mainly CO_2 and in addition a significant financial burden due to the high cost of energy [4].

A recent assessment by the International Energy Agency (IEA) regarding the status of the global building sector according to the Sustainable Development Goals (SDGs) identifies that "emissions from buildings appear to have risen again in 2018 for the second year in a row" due to "several factors, including extreme weather that raised energy demand for heating and cooling" and that "enormous potential remains untapped due to the widespread use of less-efficient technologies, a lack of effective policies and insufficient investment in sustainable buildings" [5],[6].

* Corresponding author: marinimarkaki@hmu.gr

To handle this demanding issue, the EU has set binding climate and energy targets for 2030: reducing greenhouse gas emissions by at least 40%, increasing energy efficiency by at least 32.5%, and increasing the share of renewable energy to at least 32% of EU energy use [7]. In the EU, since 2020 all new buildings are obliged to follow in all the aspects to achieve nearly zero energy consumption [8], [9].

Considering the complex framework and the large variety of buildings, a careful planning of interventions is required to pursue the energy performance objectives of public buildings promoted by the EU, accelerating renovation of the existing stock, and moving towards nearly zero energy and net zero carbon emissions buildings [10], in accordance with the National Energy Efficiency Action Plans (NEEAPs) and the NECPs.

A range of suitable energy saving measures which are already available and have applied to existing buildings, aiming to reduce energy, can be grouped into three categories:

- building envelopes (e.g., thermal insulation, suitable opening systems, passive solar systems, green roofs);
- heating and electricity loads (HVAC, electric lighting and appliances) efficiency upgrade;
- the rational use of the building (e.g., energy management);
- exploitation of RES Technologies.

A main difference between an energy-efficient building and a self-sufficient building is that in the first case a reduced energy has been succeeded, using the first three measures of the previous list. Consequently, a self-sufficient building covers the remaining energy (after the energy saving interventions) with utilization of available renewable energy sources. The most common technologies are the following:

- Photovoltaics;
- Wind turbines;
- Solar thermal heaters;
- Geothermal heat pumps.

Therefore, in order to satisfy the remaining energy demand and to maximize energy production in buildings by means of renewable sources, the extensive use of net-metering method is proposed [11]. In this paper, an analysis of extensive utilization of net metering policy instrument in all the public buildings of Crete regions is carried out. It has to be noticed that according to the current version of the Greek Legislative Framework that is related to PV net-metering, which has been updated a few months ago, batteries can also be included in PV net-metering installations, and the maximum rated power of their inverter (in kVA) cannot exceed the installed power of the PVs (in kW_p). Because our analysis is limited at the end of year 2023, we have not considered the impact of batteries in our study.

2 Case study analysis

In this section, an analysis based on measured electricity consumption data for the years 2019, 2020 and 2021 has been carried out for all the municipalities on the island of Crete.

The electrical load of the examined public buildings and premises of Region of Crete (RoC) can be categorized in, water supply (WP), street lighting (SL), technical & maintenance (TN), small offices (SO), and large premises (PR). Figure 1 depicts the annual energy demand from 2017 till 2021 of the above-mentioned types.

In 2020, consumption in all cases were less than in 2019, obviously due to COVID pandemic. It must be mentioned that energy demand has been restored in 2021, with the exception of SO and SL. The first is due to further installation of LED lights and the second due to reduced attendance in this kind of offices, even after lockdowns.

 https://doi.org/10.1051/e3sconf/202340401003

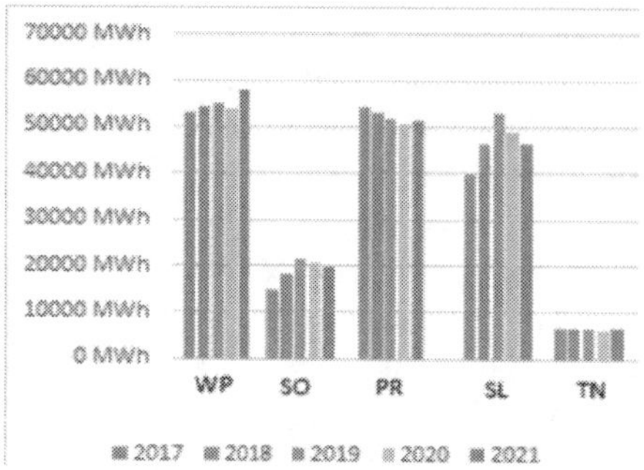

Fig. 1. Annual electric energy consumption per type.

Following, Table 1 presents the data of all the public energy meters, which count 11,636 in total for the 24 municipalities of the island. It shows the differences between demand types, the significant consumption of specific types, such as PR, and the variation of electricity prices per kWh that is equal to 95.45 €/MWh. Concluding, public buildings in RoC have the biggest share of energy (38.9%) and have the highest share in cost (50.3%), as it is shown in Figure 2.

Table 1. Consumption (in MWh) per Area and per Type of Provision for the year 2021.

Type of Service	Number of Meters	Consumption (MWh)	Total Cost (€)	Price (€/MWh)
Technical	23	7,881.94	1,234,215	156.59
Pumps	715	57,922.56	7,160,765	123.63
Small Offices	3626	20,005.98	3,071,254	148.61
Premises	982	51,358.38	11,251,547	219.08
Street Lighting	6290	46,469.76	5,791,571	124.63

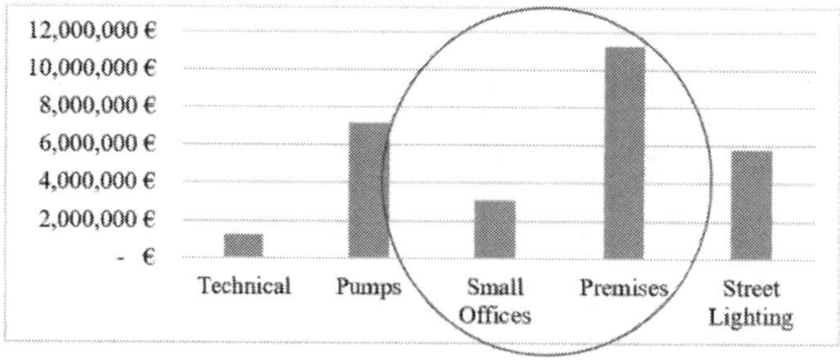

Fig. 2. Annual electricity cost per demand type in 2021.

Furthermore, Figure 3 shows that the examined large office premises of Crete's region are preferable to propose and assess any energy saving intervention, due to relatively higher cost of energy, in addition to their high demand share.

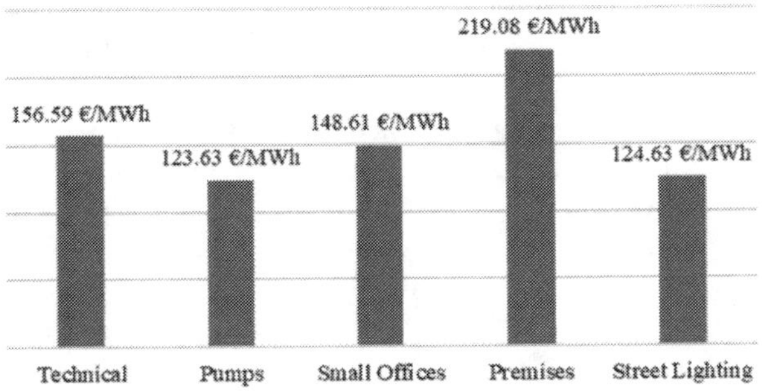

Fig. 3. Descending order for normalized types of provisions.

Additionally, two more features of the examined load profiles that can be extracted are:

- TN is the most demanding type per meter, although they represent a small percentage of them, due to its literally high electricity load demand, which is supplied through medium voltage (MV).
- The 6.290 street lightings meters represent the 54% of totally 11.636 meters, although they require less than 25% of the total annual energy demand, classifying them as the most cost-effective public services in region of Crete. This performance will be improved to its maximum, by replacing all the streetlights with LED technology.

Therefore, the high electricity demand of public buildings should be confronted, in order to improve future energy security and energy sustainability of this sector. Solar energy has been proven as the most appealing source, which can contribute to enhance buildings' sufficiency, resilience and financial completeness, targeting to the required Energy Trilemma.

3 Methodology

In this section, the potential margins for extensive use of net-metering are calculated and evaluated to cover all the corresponding electrical consumptions. Figure 4 presents the trendline of the estimated demand, which is stated as a reference, based on the years under review, for the year 2023. Considering an annual consumption of 194.93GWh in 2023, the main objective is the coverage of this demand by solar energy utilization.

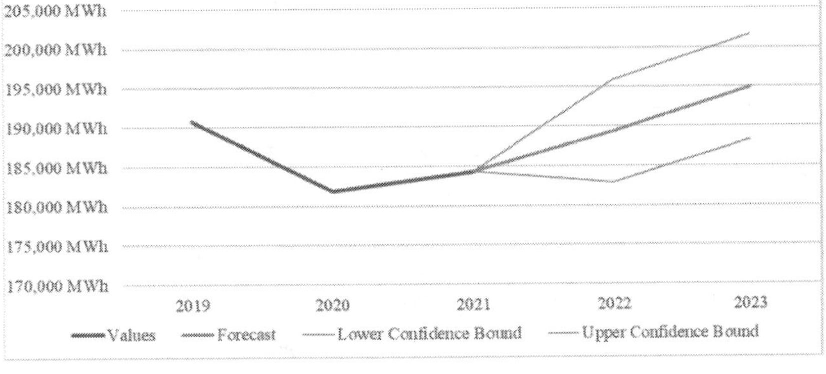

Fig. 4. Consumption forecasting.

The legislation for the installation of PV in the context of net-metering measure is taken into account. The Licensed Power Limit (LPL) of each supply is an important feature for any study of this kind.

Therefore, in order LPL of all the examined meters to be calculated, this study had used data from the corresponding payments. This was necessary to evaluate the limit of the maximum possible installed PV. The method uses the following two functions. One for the transmission systems charges:

$$LPL = \frac{TN-(C*E)}{P*Y}, \tag{1}$$

where: *TN*, is Transmission Network charge, , is the Consumption, , is the Energy Cost Factor, *P*, is the Power Cost Factor, *Y*, is the Billing Days/365.

One for distribution grid charges:

$$LPL = \frac{DN-(C*E)}{P*Y}, \tag{2}$$

where: *DN*, is Distribution Grid charge, *C*, is the Consumption, *E*, is the Energy Cost Factor, *P*, is the Power Cost Factor, *Y*, is the Billing Days/365.

It should be mentioned that the calculation of LPL for AGROT is not attainable, because the energy and power cost factors values are zero. The first assessment of the above calculation leads to the fact that PV with a total capacity of 176.93GW could be installed in the Region of Crete as max in the context of net-metering. At this point it is not examined whether the power of 176,94GW covers the energy consumption of Crete. The allocation of supplies by meter's type and level of LPL is presented in Table 2.

Additionally, Table 3 shows in parallel both the maximum required installed capacity of PV systems in order to cover all electrical load demands throughout the Region of Crete, and the calculated LPL. Particularly, the required PV power to cover all electricity consumption in 2023, is 125.76MW.

At this point it should be mentioned that it was considered a capacity factor equal to 17.7% (i.e. 1550 kWh/kW), being adjusted by the European GIS [12]. In addition, it should be noted that these PV are subjected to the limitation of each substation available capacity of each local distribution grid.

Table 2. Allocation of Supplies per type and capability.

Tariffs	Agreed Power (kVA)	No Supllies	PV Power Limit (kW)	Total (kW)
G1	8	9	72	127
	15	2	30	
	25	1	25	
G21	8	2,174	17,392	45,697
	15	672	10,080	
	25	729	18,225	
G22	35	368	12,880	75,055
	55	273	15,015	
	85	142	12,070	
	135	134	18,090	
	250	68	17,000	
G23	8	1	8	1,033
	15	1	15	
	25	15	375	
	35	11	385	
	250	1	250	
FOP	8	5,923	47,384	55,024
	15	226	3,390	
	25	143	3,575	
	35	13	455	
	55	4	220	
Totals:		**10,910**	**176,936**	**176,936**

Table 3. Consumption (kWh) and PV Power Limits per Regional Unit for the year 2023.

Regional Unit	Consumption (kWh) 2021	Predicted Consumption (kWh) 2023	PV POWER (kW)	Agreed Power (A.P.)(kVA)	PV Power Limit (kW)
Heraklion	84,778,777	88,030,185	56,793.667	67,060.000	67,060.000
Chania	47,909,722	48,310,735	31,168.216	49,008.000	49,008.000
Rethymno	23,127,555	25,030,990	16,149.026	32,213.000	32,213.000
Lasithi	28,484,157	31,587,414	20,378.976	28,655.000	28,655.000
Total:	**184,300,211**	**192,959,323**	**124,489.886**	**176,936.000**	**176,936.000**

Figure 5 shows the geographical dispersion of the proposed installed capacity per regional units of Crete Island. In the rest of the paper Heraklion will be referred to as Area 1, Chania as Area 2, Rethymno as Area 3 and Lasithi as Area 4.

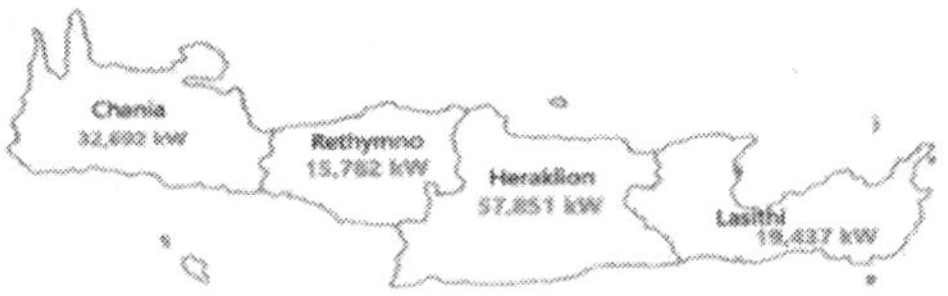

Fig. 5. Geographical dispersion of the proposed PV per area.

The same methodology is carried out for the calculation and evaluation of the installation margins potential, in the geographical subunits of each Area separately. In this way, a more accurate and detailed analysis is achieved for the determination of the maximum installed capacity limits of PV for the energy coverage of the respective consumptions, in accordance with the specific energy demands. According to the provided data, four forecasts were made, as shown in the following figures (Figure 6 to Figure 9)

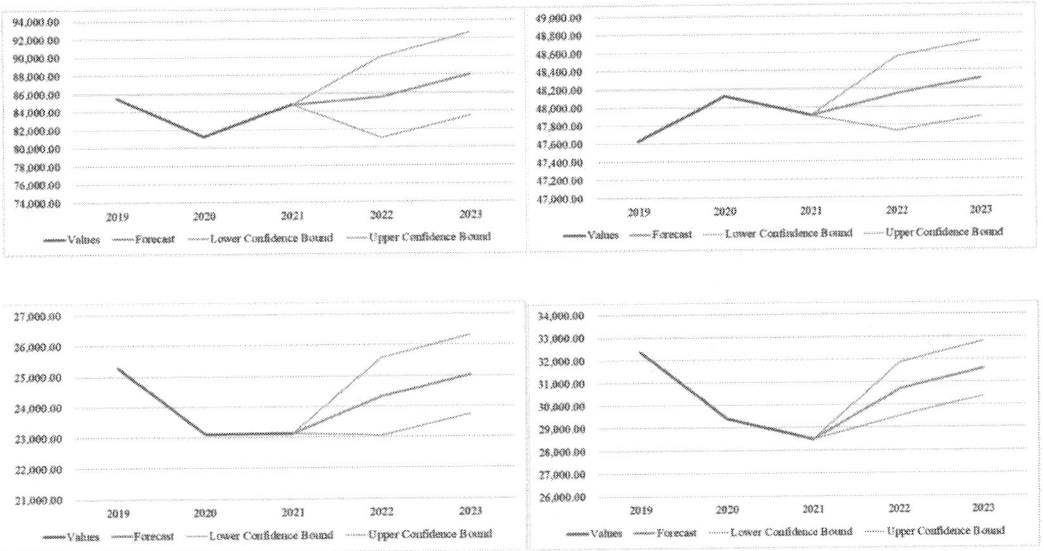

Fig. 6. Demand forecast for (a) Area 1, (b) Area 2, (c) Area 3 and (d) Area 4.

At this point, it is noted that the predicted consumption of the year 2023 according to the first analysis (made for the whole of Crete) was 194.93GWh while the current analysis (per Regional Unit of Crete) it results to 192.96GWh. There is a difference of 1.97GWh, concluding that the more accurate a study is, the more approximate and realistic our results will be.

Subsequently, according to the carried-out calculations, four estimated values are derived as the total maximum power of PV systems needed in the context of net-metering. More specifically, about Areas 1, 2, 3 and 4, the total power of PV systems that could be installed are 67.06MW, 49.01MW, 32.21MW and 28.65MW, respectively. Cumulatively, the total power of PV adds to 176.93MW. The following Tables 4 to 7 present the aforementioned results.

Table 4. Allocation of Supplies per A.P. and Capability for Area 1.

Tariffs	Agreed Power (kVA)	No Supplies	PV Power Limit (kW)	Total (kW)
G1	8	5	40	40
	8	842	6,736	
G21	15	331	4,965	19,976
	25	331	8,275	
	35	155	5,425	
	55	138	7,590	
G22	85	59	5,015	30,855
	135	45	6,075	
	250	27	6,750	
	15	1	15	
G23	25	3	75	160
	35	2	70	
	8	1,793	14,344	
FOP	15	29	435	16,029
	25	43	1,075	
	35	5	175	
Totals:		3,809	67,060	67,060

Table 5. Allocation of Supplies per A.P. and Capability for Area 2.

Tariffs	Agreed Power (kVA)	No Supplies	PV Power Limit (kW)	Total (kW)
G1	8	4	32	57
	25	1	25	
G21	8	506	4,048	
	15	80	1,200	9,223
	25	159	3,975	
	35	102	3,570	
G22	55	63	3,465	
	85	44	3,740	20,425
	135	40	5,400	
	250	17	4,250	
G23	25	2	50	50
	8	2,071	16,568	
	15	88	1,320	
FOP	25	41	1,025	19,253
	35	5	175	
	55	3	165	
Totals:		3,223	49,008	49,008

Table 6. Allocation of Supplies per A.P. and Capability for Area 3.

Tariffs	Agreed Power (kVA)	No Supplies	PV Power Limit (kW)	Total (kW)
G1	15	2	30	30
G21	8	551	4,408	10,108
	15	150	2,250	
	25	138	3,450	
G22	35	60	2,100	13,350
	55	40	2,200	
	85	24	2,040	
	135	26	3,510	
	250	14	3,500	
G23	25	9	225	790
	35	9	315	
	250	1	250	
FOP	8	915	7,320	7,935
	15	8	120	
	25	17	425	
	35	2	70	
Totals:		1,966	32,213	32,213

Table 7. Allocation of Supplies per A.P. and Capability for Area 4.

Tariffs	Agreed Power (kVA)	No Supplies	PV Power Limit (kW)	Total (kW)
G21	8	275	2,200	6,390
	15	111	1,665	
	25	101	2,525	
G22	35	51	1,785	10,425
	55	32	1,760	
	85	15	1,275	
	135	23	3,105	
	250	10	2,500	
G23	8	1	8	33
	25	1	25	
FOP	8	1,144	9,152	11,807
	15	101	1,515	
	25	42	1,050	
	35	1	35	
	55	1	55	
Totals:		1,909	28,655	28,655

Table 8 shows in parallel both the maximum required capacity of PV systems in order to cover all the electrical demands, and the limits due to LPL. In total, the required PV capacity, which is necessary to cover all energy consumption amounts to 124.49MW.

Table 8. Consumption and PV Power per Area in 2023.

Regional Unit	Consumption (kWh) 2021	Predicted Consumption (kWh) 2023	PV POWER (kW)	Agreed Power (A.P.)(kVA)	PV Power Limit (kW)
Heraklion	84,778,777	88,030,185	56,793.667	67,060.000	67,060.000
Chania	47,909,722	48,310,735	31,168.216	49,008.000	49,008.000
Rethymno	23,127,555	25,030,990	16,149.026	32,213.000	32,213.000
Lasithi	28,484,157	31,587,414	20,378.976	28,655.000	28,655.000
Total:	**184,300,211**	**192,959,323**	**124,489.886**	**176,936.000**	**176,936.000**

At this point similarly mentioned above, it should be mentioned that it was considered a capacity factor equal to 17.7% (i.e. 1550 kWh/kW). In addition, it should be noted that these PV installations are subject to the limitation of each substation of the local distribution grids.

4 Results

Generally, to evaluate an investment, various indices must be considered that determine whether it is profitable or not over a predetermined period of time. Such indices are Net Present Value (NPV), Internal Rate of Return (IRR) and Payback Period (PBP).

The CAPEX of PV is considered equal to 1,000€ per kWp. For project A, an investment for the total coverage of the predicted consumption in 2023 (i.e. 192,9GWh), a total of 124,5MWp in PV is required, which amounts to a total CAPEX equal to 124,490,000€. At the same time, for project B, an investment for the total coverage of the projected consumptions in 2023 for each category of tariffs, the total costs for each investment are presented based on demand type in Table 9.

Table 9. Investment Cost per public service type.

Category	Consumption 2023 (kWh)	kWp 2023	Investment Cost (€)
Agricultural	60,643,974.98	39,125	39,125,000.00
Industrial	8,252,273.44	5,324	5,324,000.00
Residential	45,036.01	29	29,000.00
Commercial	75,364,752.64	48,623	48,623,000.00
Street Lighting	48,653,285.94	31,389	31,389,000.00
Total	**192,959,323.00**	**124,490**	**124,490,000.00**

Mainly, an investment is a capital commitment that is expected to bring additional capital to the investor in a predetermined period of time. More specifically, the investment consists of a sequence of Free Cash Flow (FCF). Cash flows are divided into cash inflows and outflows, the difference between them constitutes the Free Cash Flow (FCF).

$$FCF_i = Cash\ inflows_i - Cash\ outflows_i. \quad (3)$$

Also, in order to calculate the FCF, an average Degradation Rate of photovoltaics has been applied and the lifetime of the panels are considered to be twenty years. The degradation rate is the annual decrease of a photovoltaic panel's efficiency, due to overheating of photovoltaic cells, resulting to the alteration of the structure of their materials. Depending on the environment's temperature or the area of installation of the panels, the degradation rate ranges from 0.5% and above [13]. In the current application, the degradation rate has been set at 0.5%.

Project A:

This investment project, as mentioned initially, involves the coverage of public energy consumptions of the entire Region of Crete for 2023. In addition, in order to calculate the cash flows, a single cost price of MWh has been set at 157,56 €/MWh, which is the average of the normalized prices of the tariffs. As mentioned before, for an investment to be considered profitable, some rating indicators should be taken into account, as it is shown in Table 10.

https://doi.org/10.1051/e3sconf/202340401003

Table 10. Financial assessment of Project A.

Year	Production (MWh) with Degradation Rate applied	Cash Flows	Annual Present Values (PV)
		-124,490,000.00 €	-124,490,000.00 €
2023	192,959.50 MWh	30,401,841.22 €	30,100,832.89 €
2024	191,994.70 MWh	30,249,832.02 €	29,653,790.82 €
2025	191,034.73 MWh	30,098,582.86 €	29,213,387.99 €
2026	190,079.56 MWh	29,948,089.94 €	28,779,525.79 €
2027	189,129.16 MWh	29,798,349.49 €	28,352,107.09 €
2028	188,183.51 MWh	29,649,357.74 €	27,931,036.19 €
2029	187,242.59 MWh	29,501,110.96 €	27,516,218.82 €
2030	186,306.38 MWh	29,353,605.40 €	27,107,562.11 €
2031	185,374.85 MWh	29,206,837.37 €	26,704,974.55 €
2032	184,447.98 MWh	29,060,803.19 €	26,308,366.02 €
2033	183,525.74 MWh	28,915,499.17 €	25,917,647.71 €
2034	182,608.11 MWh	28,770,921.68 €	25,532,732.15 €
2035	181,695.07 MWh	28,627,067.07 €	25,153,533.16 €
2036	180,786.59 MWh	28,483,931.73 €	24,779,965.83 €
2037	179,882.66 MWh	28,341,512.07 €	24,411,946.54 €
2038	178,983.24 MWh	28,199,804.51 €	24,049,392.88 €
2039	178,088.33 MWh	28,058,805.49 €	23,692,223.68 €
2040	177,197.89 MWh	27,918,511.46 €	23,340,358.97 €
2041	176,311.90 MWh	27,778,918.91 €	22,993,719.97 €
2042	175,430.34 MWh	27,640,024.31 €	22,652,229.08 €
Opportunity Cost (i) :		1.00%	
Net Present Value (NPV) :		399,701,552.25 €	
Iternal Rate of Return (IRR) :		23.60%	
Payback Period (PBP) :		4.24 years	

Project B:

As already mentioned, this investment project targets the coverage of the projected consumptions in 2023 for each demand type separately. Therefore, in this project there is one investment per type, thus giving the investor the choice. In addition, in order to calculate the cash flows, an average normalized price of €/MWh has been determined for each demand type (category of tariffs). Concluding, Tables 11 to 14 below, present the assessment of each investment of project B.

Table 11. Indicators of Project B for TN.

Year	Production (MWh) with Degradation Rate applied	Cash Flows	Annual Present Values (PV)
		-48,623,000.00 €	-48,623,000.00 €
2023	75,365.65 MWh	12,547,153.78 €	12,422,924.54 €
2024	74,988.82 MWh	12,484,418.01 €	12,238,425.66 €
2025	74,613.88 MWh	12,421,995.92 €	12,056,666.86 €
2026	74,240.81 MWh	12,359,885.94 €	11,877,607.45 €
2027	73,869.60 MWh	12,298,086.51 €	11,701,207.34 €
2028	73,500.26 MWh	12,236,596.08 €	11,527,427.03 €
2029	73,132.75 MWh	12,175,413.10 €	11,356,227.62 €
2030	72,767.09 MWh	12,114,536.03 €	11,187,570.78 €
2031	72,403.26 MWh	12,053,963.35 €	11,021,418.73 €
2032	72,041.24 MWh	11,993,693.54 €	10,857,734.30 €
2033	71,681.03 MWh	11,933,725.07 €	10,696,480.82 €
2034	71,322.63 MWh	11,874,056.44 €	10,537,622.19 €
2035	70,966.01 MWh	11,814,686.16 €	10,381,122.85 €
2036	70,611.18 MWh	11,755,612.73 €	10,226,947.76 €
2037	70,258.13 MWh	11,696,834.67 €	10,075,062.40 €
2038	69,906.84 MWh	11,638,350.49 €	9,925,432.76 €
2039	69,557.30 MWh	11,580,158.74 €	9,778,025.34 €
2040	69,209.52 MWh	11,522,257.95 €	9,632,807.14 €
2041	68,863.47 MWh	11,464,646.66 €	9,489,745.65 €
2042	68,519.15 MWh	11,407,323.42 €	9,348,808.83 €
Opportunity Cost (i) :		1.00%	
Net Present Value (NPV) :		167,716,266.05 €	
Iternal Rate of Return (IRR) :		25.04%	
Payback Period (PBP) :		4.00 years	

Table 12. Investment indices of Project B for SP and LP.

Year	Production (MWh) with Degradation Rate applied	Cash Flows	Annual Present Values (PV)
		-29,000.00 €	-29,000.00 €
2023	44.95 MWh	7,663.91 €	7,588.03 €
2024	44.73 MWh	7,625.60 €	7,475.34 €
2025	44.50 MWh	7,587.47 €	7,364.32 €
2026	44.28 MWh	7,549.53 €	7,254.95 €
2027	44.06 MWh	7,511.78 €	7,147.20 €
2028	43.84 MWh	7,474.22 €	7,041.06 €
2029	43.62 MWh	7,436.85 €	6,936.49 €
2030	43.40 MWh	7,399.67 €	6,833.47 €
2031	43.18 MWh	7,362.67 €	6,731.98 €
2032	42.97 MWh	7,325.86 €	6,632.00 €
2033	42.75 MWh	7,289.23 €	6,533.51 €
2034	42.54 MWh	7,252.78 €	6,436.47 €
2035	42.33 MWh	7,216.52 €	6,340.88 €
2036	42.11 MWh	7,180.43 €	6,246.71 €
2037	41.90 MWh	7,144.53 €	6,153.94 €
2038	41.69 MWh	7,108.81 €	6,062.54 €
2039	41.49 MWh	7,073.27 €	5,972.51 €
2040	41.28 MWh	7,037.90 €	5,883.81 €
2041	41.07 MWh	7,002.71 €	5,796.42 €
2042	40.87 MWh	6,967.70 €	5,710.34 €
Opportunity Cost (i) :		**1.00%**	
Net Present Value (NPV) :		**103,141.98 €**	
Iternal Rate of Return (IRR) :		**25.68%**	
Payback Period (PBP) :		**3.91 years**	

Table 13. Investment indices of Project B for WP.

Year	Production (MWh) with Degradation Rate applied	Cash Flows	Annual Present Values (PV)
		-5,324,000.00 €	-5,324,000.00 €
2023	8,252.20 MWh	1,376,143.20 €	1,362,518.02 €
2024	8,210.94 MWh	1,369,262.49 €	1,342,282.61 €
2025	8,169.88 MWh	1,362,416.17 €	1,322,347.72 €
2026	8,129.03 MWh	1,355,604.09 €	1,302,708.89 €
2027	8,088.39 MWh	1,348,826.07 €	1,283,361.73 €
2028	8,047.95 MWh	1,342,081.94 €	1,264,301.90 €
2029	8,007.71 MWh	1,335,371.53 €	1,245,525.14 €
2030	7,967.67 MWh	1,328,694.67 €	1,227,027.24 €
2031	7,927.83 MWh	1,322,051.20 €	1,208,804.06 €
2032	7,888.19 MWh	1,315,440.95 €	1,190,851.53 €
2033	7,848.75 MWh	1,308,863.74 €	1,173,165.61 €
2034	7,809.51 MWh	1,302,319.42 €	1,155,742.36 €
2035	7,770.46 MWh	1,295,807.82 €	1,138,577.87 €
2036	7,731.61 MWh	1,289,328.79 €	1,121,668.30 €
2037	7,692.95 MWh	1,282,882.14 €	1,105,009.86 €
2038	7,654.48 MWh	1,276,467.73 €	1,088,598.82 €
2039	7,616.21 MWh	1,270,085.39 €	1,072,431.51 €
2040	7,578.13 MWh	1,263,734.97 €	1,056,504.31 €
2041	7,540.24 MWh	1,257,416.29 €	1,040,813.65 €
2042	7,502.54 MWh	1,251,129.21 €	1,025,356.03 €
Opportunity Cost (i) :		**1.00%**	
Net Present Value (NPV) :		**18,403,597.16 €**	
Iternal Rate of Return (IRR) :		**25.08%**	
Payback Period (PBP) :		**4.00 years**	

 https://doi.org/10.1051/e3sconf/202340401003

Table 14. Investment Indices of Project B for SL.

Year	Production (MWh) with Degradation Rate applied	Cash Flows	Annual Present Values (PV)
		-31,389,000.00 €	-31,389,000.00 €
2023	48,652.95 MWh	6,063,673.57 €	6,003,637.20 €
2024	48,409.69 MWh	6,033,355.20 €	5,914,474.27 €
2025	48,167.64 MWh	6,003,188.43 €	5,826,635.55 €
2026	47,926.80 MWh	5,973,172.49 €	5,740,101.35 €
2027	47,687.16 MWh	5,943,306.62 €	5,654,852.32 €
2028	47,448.73 MWh	5,913,590.09 €	5,570,869.37 €
2029	47,211.49 MWh	5,884,022.14 €	5,488,133.68 €
2030	46,975.43 MWh	5,854,602.03 €	5,406,626.75 €
2031	46,740.55 MWh	5,825,329.02 €	5,326,330.31 €
2032	46,506.85 MWh	5,796,202.37 €	5,247,226.40 €
2033	46,274.31 MWh	5,767,221.36 €	5,169,297.29 €
2034	46,042.94 MWh	5,738,385.26 €	5,092,525.55 €
2035	45,812.73 MWh	5,709,693.33 €	5,016,893.98 €
2036	45,583.66 MWh	5,681,144.86 €	4,942,385.66 €
2037	45,355.75 MWh	5,652,739.14 €	4,868,983.89 €
2038	45,128.97 MWh	5,624,475.44 €	4,796,672.25 €
2039	44,903.32 MWh	5,596,353.07 €	4,725,434.54 €
2040	44,678.81 MWh	5,568,371.30 €	4,655,254.82 €
2041	44,455.41 MWh	5,540,529.44 €	4,586,117.37 €
2042	44,233.13 MWh	5,512,826.80 €	4,518,006.72 €
Opportunity Cost (i) :		**1.00%**	
Net Present Value (NPV) :		**73,161,459.27 €**	
Iternal Rate of Return (IRR) :		**18.20%**	
Payback Period (PBP) :		**5.40 years**	

The investments with the highest IRR, are the most profitable and the most likely to be chosen. So, the most profitable investment is the demand type of public buildings (both SP and LP) of project B.

5 Conclusions

Extensive use of net-metering method in public buildings is a promising solution towards the energy transition and additionally an attractive investment.

In the examined public facilities of Crete's region, LP and SP are the most preferable to introduce electricity saving and/or production measures, due to comparatively high electricity cost. Considering a specific demand forecasting for 2023 which is 194.93GWh, required PV capacity to cover it all should be 125MW, approximately.

Future research will focus on evaluating the total impact of more specific types of public buildings (schools, sport facilities, campuses) towards the energy transition of the island, as well as on studying the impact of batteries in net-metering installations.

This work received financial support from the project "Enhancing resilience of Cretan power system using distributed energy resources (CResDER)" (Proposal ID: 03698) financed by the Hellenic Foundation for Research and Innovation (H.F.R.I.) under the Action "2nd Call for H.F.R.I. Research Projects to support Faculty Members and Researchers".

References

1. Y. Xing, N. Hewitt, P. Griffiths, Ren. Sust. En. Rev. **15**, 3229-3236 (2011)
2. M.González-Torres, L.Pérez-Lombard, J. F. Coronel, I. R. Maestre, D. Yanc, En. Rep. **8**, 626-637 (2022)
3. European Commission, Clean energy for all Europeans package (2019), available athttps://ec.europa.eu/energy/topics/energy-strategy/clean-energy-all-europeans_en (accessed on 13 Feb 2023)
4. European Commission, Energy Performance of Buildings Directive (2018), available at https://ec.europa.eu/energy/topics/energy-efficiency/energy-efficient-buildings/energy-performance-buildings-directive_en (accessed on 13 Feb 2023)

5. Directive 2010/31/EU of the European Parliament and of the Council of 19 May 2010 on the energy performance of buildings (2010), available at https://eur-lex.europa.eu/eli/dir/2010/31/oj (2010) (accessed on 13 Feb 2023)
6. A. Kumar, H. C. Borah, U. Goswami, (2021). Grid Connected Rooftop Solar Photo-Voltaic Power Plants Programme in India: *A Review of State Solar Rooftop Policies, Net Metering Regulations, Challenges and Way Forward*, advances in Smart Communication Technology and Information Processing, OPTRONIX, India (2021)
7. International Energy Agency, Tracking buildings 2019 (2019).
8. European Commission, Official Journal of European Union 13-35 (2010)
9. R. Dufo-López, J. L. Bernal-Agustín, En. **84**, 684-694 (2015)
10. C. Cooremans, A. Schönenberger, J. Cl. Prod. **230**, 264-275 (2019)
11. W. U. Rehman, A. R. Bhatti, A. B. Awan, I. A. Sajjad, A. A. Khan, R. Bo, S. S. Haroon, S. Amin, I. Tlili, O. Oboreh-Snapps, IEEE Access **8**, 170364-170388 (2020)
12. Photovoltaic Geographical Information System, avalable online: https://re.jrc.ec.europa.eu/pvg_tools/en/ (accessed on 13 Feb 2023)
13. NREL Transforming ENERGY, available online: https://www.nrel.gov/state-local-tribal/blog/posts/stat-faqs-part2-lifetime-of-pv-panels.html (accessed on 13 Feb 2023)

Design and calculation of an environmentally friendly carbon-free hybrid plant based on a microgas turbine and a solid oxide fuel cell

Ivan Beloev[1], *Antonina* Filimonova[2], *Andrey* Chichirov[2*], *Artem* Vinogradov[2] and *Iliya* Iliev[3]

[1]University of Ruse, Department of Transport, 7004 Ruse, Bulgaria
[2]Kazan State Power Engineering University, Department of Chemistry and Hydrogen Energy, 420000 Kazan, Russia
[3]University of Ruse, Department of Heat, Hydraulics and Environmental Engineering, 7004 Ruse, Bulgaria

Abstract. This is an overview of a hybrid power plant design and pre-design analysis, including a microgas turbine with heat recovery, a high-temperature fuel cell, and a carbon dioxide capture system. A hybrid installation model is presented, taking into account the compatibility and technological limitations of the main components. The material and heat balance calculation of a hybrid power plant is performed depending on the input parameters under partial load conditions. In order to create a decarbonized highly efficient energy production process and in connection with the need to minimize the negative impact of carbon dioxide on the environment, the article presents the developed technologies for carbon dioxide utilization and a carbon adsorption unit as a hybrid power plant part. The hybrid power plant is a carbon-free mini thermal power plant with integrated electricity, steam, and hot water generation and more than 90% total efficiency.

1 Introduction

Due to the growing demand for electricity and the downward trend in the use of fossil fuels, most countries are developing highly efficient and cost-effective energy systems. Fuel cells are one of the most promising sources of energy due to their high level of efficiency and low emission of environmentally harmful gases. To achieve higher efficiency, a solid oxide fuel cell (SOFC) stack can be coupled to a gas turbine (GT) thermodynamic cycle. In this case, the calculated electrical efficiency (EE) becomes higher than 60% and corresponds to large generating power plants (PP). In addition, it contributes to the development of fuel cell production and the expansion of the gas turbine industry [1-3].

However, despite the advantages of current SOFC-GT technology, many technical barriers must be overcome in order to successfully develop a highly efficient hybrid system. These difficulties are especially evident in large-scale power systems, because a huge

* Corresponding author: aachichirova@mail.ru

number of installations must work stably together, and the SOFC power generation system and the gas turbine must be integrated to ensure safe and stable operation without any structural and physical damage of the stack. As a rule, the mode of operation of a stand-alone gas turbine is dynamic, while the mode of operation of an SOFC power generation system is static [4].
Numerous different configurations of SOFC-GT hybrid systems [5, 6] have been proposed in the literature to improve EE and/or reduce the investment costs.

The design depends from parameters and should be made a choice of SOFC-GT system layout between working temperature and pressure in SOFC, fuel type and features of the other processing subsystem (steam reforming: direct/indirect; internal/external; autothermal reforming, partial oxidation, etc.), steam for the reforming process: recirculation of anode exhaust gases or external steam generator, type of Brayton cycle: main, with intercooling and/or reheating, etc. The layout features of these SOFC-GT systems are presented in Table 1.

The main structural element of the hybrid PP is a solid oxide fuel cell. SOFCs were considered as one of the most promising energy conversion technologies. The main feature of this installation is the operation at high operating temperatures, demonstrating ultra-high electrical and thermal efficiency, regardless of the size of the system. The main types of SOFCs used are tubular, microtubular and planar [7].

The other investigation of design and type of interconnections between the modules of a hybrid system have a important impact on the achievement of performance targets based on the results of numerical simulation of the proposed circuit solutions [8-10].

Table 1. Features, advantages and disadvantages of SOFC hybrid plant cycles.

<table>
<tr><th>Cycle</th><th>Features - advantages</th><th>Disadvantages</th></tr>
<tr><td>Tubular SOFC</td><td>Possibility of using a thin layer of electrolyte; fast response to load changes.</td><td>It is more difficult to optimize the current collection, especially the current collection on the anode side of the tube.</td></tr>
<tr><td>Micro-tubular SOFC</td><td>High volume current density, good thermal cycle tolerance, easy sealing between fuel and oxidant streams, low capital cost, small size</td><td>Higher production cost, larger installation dimensions.</td></tr>
<tr><td>Planar SOFC</td><td>High power density, simplicity of the technological process, compact installation.</td><td>The system is characterized by a high degree of inertia.</td></tr>
<tr><td rowspan="5">Internal reforming</td><td colspan="2">Direct internal reforming</td></tr>
<tr><td>Natural gas is converted into a hydrogen-rich mixture inside the anode compartment; simplicity of the system and low capital costs.</td><td>The anode compartment needs a catalyst for the methane steam reforming reaction; risk of carbon deposition.</td></tr>
<tr><td colspan="2">Anode gases recirculation</td></tr>
<tr><td>Recirculation of anode exhaust gases to the fuel processing subsystem; simplicity and low capital costs; high conversion efficiency.</td><td>SOFC temperature profiles depend on gas flow distribution, cell operating parameters and the reforming process; integrated launch.</td></tr>
<tr><td colspan="2">External steam generator</td></tr>
</table>

https://doi.org/10.1051/e3sconf/202340401004

	The produced steam supports the steam reforming reaction using exhaust heat; easy control of production and use of steam for thermal purposes.	High capital costs and system complexity.
SOFC/GT external reforming	SOFC exhaust gases are used for external reforming; use of more complex fuels (biogas, synthesis gas, liquids, etc.)	Integrated management of thermal regimes; cost of additional fuel; lower efficiency; high capital costs.
Hybrid trigeneration cycle	Use of various cycles (for example, the organic Rankine cycle); high temperature of the exhaust gases allows ultra-high conversion efficiency to be achieved.	High capital costs and system complexity; available for high power installations.
Hybrid SOFC/GT with air or exhaust gases recirculation	Recuperative heat exchanger	
	Preheating of the air entering the cathode compartment of the fuel cell with the help of exhaust gases from the combustion chamber.	Reduced turbine inlet temperature as the combustion chamber outlet is cooled in a recuperative heat exchanger before entering the GT.
	Exhaust gas recirculation	
	A part of the output stream from the combustion chamber is recirculated to the cathode inlet, raising the temperature at the cathode inlet.	With an increase in the temperature at the turbine inlet, the efficiency decreases more than in a cycle with a recuperative heat exchanger.
System pressure	Atmospheric	
	Operational safety and reliability	High price
	Increased	
	High EE	Control complexity

The high operating temperature of the SOFC improves the overall efficiency of the hybrid system and can be used as a heat source for various purposes. This technology is especially attractive for operation as a mini-CHPP.

For social facilities and industrial enterprises, the average electrical consumption is less than 100 kW/facility, and below 10 kW for the service sector. It should also be noted that the ratio of consumed electrical and thermal energy is usually 40/60%. Decentralized natural gas PP can provide combined heat and power generation (CHPP) and flexibility due to localized generation. Solid oxide fuel cells (SOFCs) can cogenerate electricity and heat from a wide range of RES (natural gas, synthetic natural gas or biogas). SOFC heat can be used for example, process heat (industrial purposes), and for heating and hot water [11].

Another important issue in the production of electricity using modern technologies is the reduction of greenhouse gas emissions. For fossil or RES such as synthetic natural gas, carbon capture and storage (CCS) and carbon capture and utilization (CCU) technologies should be implemented. Many researchers are studying various CCS technologies, as well as their implementation in various industrial and energy applications [12-15]. CO2 capture technologies for SOFC-GT systems can be divided into three categories depending on the process stage at which they are applied:

- Pretreatment – separation of carbon dioxide in the fuel before its conversion (for example, separation of CO2 from H2 or CH4).
- Oxygen combustion – combustion of fuel using pure O2 (without N2) to create a high purity CO2 stream.
- Final purification – separation of CO2 from the flue gas mixture after fuel conversion.

In SOFC hybrid systems, the composition of the exhaust gases depends on the operating conditions, type of fuel used, type of reforming, fuel utilization rate, SOFC type, hybrid system layout, operating temperature, pressure, etc. Exhaust gases from a hybrid SOFC-GT system running on methane contain less than 1% hydrogen, 5% CO, 30% CO2 and about 60% water vapor. Therefore, it is necessary to capture the carbon dioxide released during the operation of the hybrid system to ensure the decarbonization of the energy production process.

Reviews and theoretical studies of SOFC-GT hybrid systems and carbon dioxide capture technologies are widely presented in the scientific literature. However, studies integrating SOFC-GT systems with a carbon capture unit for a decarbonized power and heat generation cycle are not widely represented. Due to the relevance of the topic, the aim of the work was to develop a technological scheme for a hybrid PP, including a gas microturbine with heat recovery, a high-temperature fuel cell and a carbon dioxide capture system in the format of a mini-CHPP.

2 Materials and methods

In the study, a PP system with a 30 kW SOFC and a 30 kW gas microturbine with a pre-reformer for complex fuels (hydrocarbons, biofuels, biomass) was designed. Anode-supported planar SOFC stack is a Chinese manufacturer's model. According to the proposed scheme, the calculation of the technical and economic characteristics and the average cost of the equipment used was carried out.

Air disposal (AD), fuel disposal (FD), oxygen to carbon (O/C) or steam to carbon (S/C) ratios and recirculation ratio (r) are characteristic parameters that need to be monitored and adjusted in order to keep SOFC in optimal operating condition without the risk of increased degradation or irreversible damage to the fuel cells [6].

$$\frac{O}{C}=\frac{nO}{nC}=\frac{xH_2O+xCO+2xCO_2}{\sum_N N\times xC_nH_m+xCO+xCO_2}; \quad (1)$$

$$\frac{S}{C}=\frac{nH_2O}{nC}=\frac{xH_2O}{\sum_N N\times xC_nH_m+xCO+xCO_2}; \quad (2)$$

$$FD=\frac{n_{conv.fuel}}{n_{fuel}}; \quad (3)$$

$$AD=\frac{I\times N_{cells}}{4\times F\times n_{air.inl.}\times xO_{2inl.}}, \quad (4)$$

where n is the molar flow rate, mol/s; N is the number of cells; x – concentration, %; F is the Faraday number; I is the stack current.

The theoretical EE of the SOFC PP can be calculate by:

$$\eta_t=\frac{P_{net}}{P_{tp}}, \quad (5)$$

where:

P_{net} – net output electric power, W;

P_{tp} – total power of the energy supply to the system,

$$P_{tp} = LCV \times m, \tag{6}$$

where:

LCV - lower calorific value of fuel, MJ/kg

m - fuel consumption, kg/h

The energy balance of a hybrid PP is described by the following equation:

$$m_{air} \times h_{air} + m_{f,SOFC} \times \beta_f \times L_f + \int \frac{dQ_{CC}}{dt} - m_{EG} \times h_{EG} - \int \frac{dQ_{loss}}{dt} - P_{SOFC,DC} - P_{GT} = 0, \tag{7}$$

where: the mass flow is denoted with m [kg/h]; the enthalpy is h [J/kg]; β_f is SOFC fuel utilization factor; L_f [MJ/kg] is lower calorific value of fuel; P [kW] is the power; f is the fuel; CC is combustion chamber; GT is gas turbine; DC is direct current; EG is exhaust gases.

With the help of equation (8) we can estimated the EE of a hybrid PP:

$$\eta_{el} = \frac{P_{net}}{P_{tp}}, \tag{8}$$

where P_{tp} [W] is the total power of energy supply and P_{net} [W] is net output power, where:

$$P_{net} = \eta_{inv} \times P_{SOFC,DC} + \eta_{gen} \times P_{GT}, \tag{9}$$

η_{inv}, η_{gen} – efficiency of inverter and generator.

$$P_{total} = m_{f,SOFC} \times \beta_f \times L_f + P_{CC}, \tag{10}$$

3 Results

The hybrid plant of the SOFC/GT scheme is shown in Figure 1 and Figure 2.

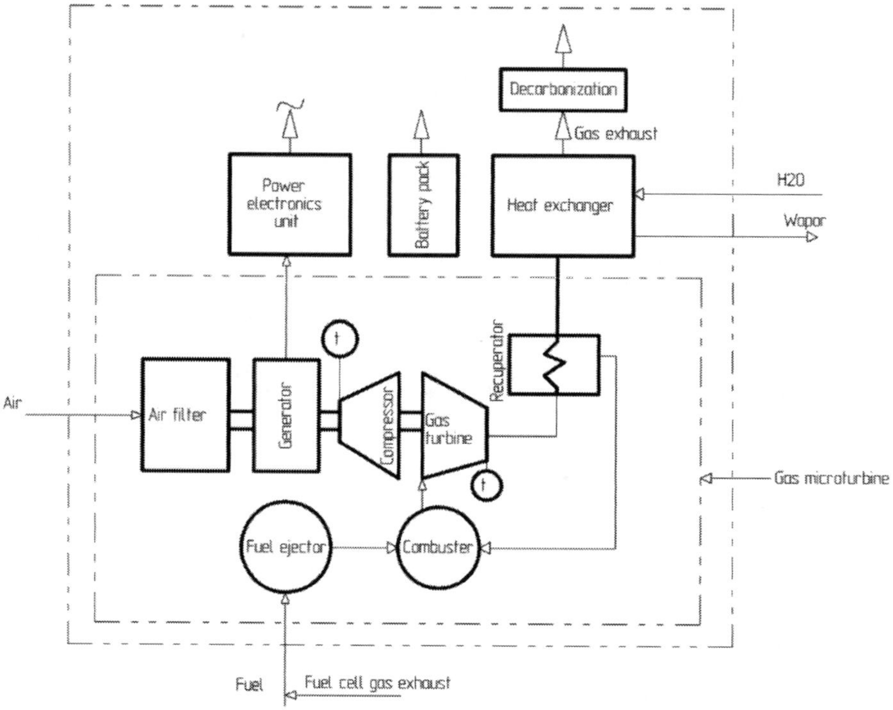

Fig. 1. Block diagram of a hybrid PP. Gas microturbine block.

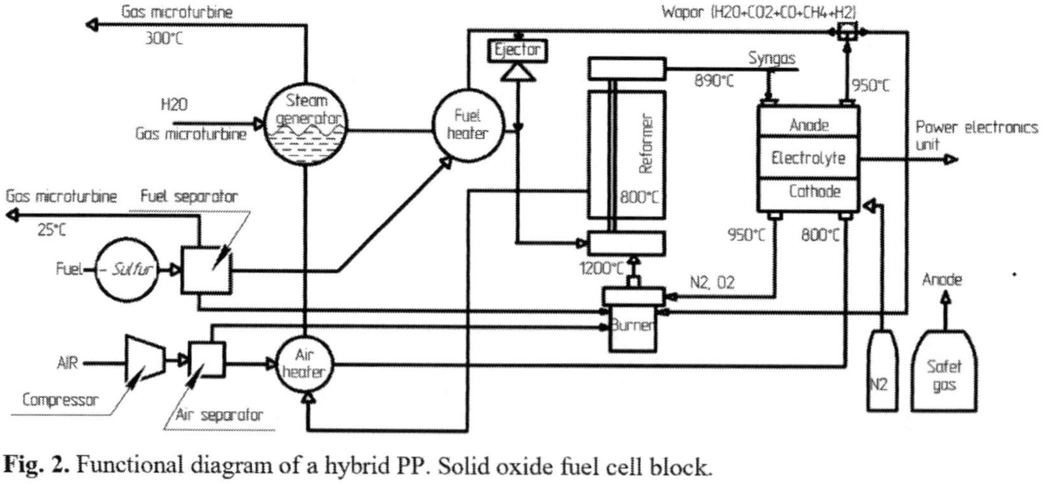

Fig. 2. Functional diagram of a hybrid PP. Solid oxide fuel cell block.

Fuel (methane) and air enter the distribution module and are fed to the gas microturbine and the SOFC. In the case of using complex hydrocarbons, industrial wastes, biofuels, natural gas, etc., the fuel is subjected to desulfurization and preliminary reforming. The fuel passes through the steam reforming stage and is supplied as synthesis gas to the SOFC anode. Heated air is supplied to the SOFC cathode. In SOFC, an electrochemical reaction takes place, as a result of which an electric current and heat are generated. Exhaust gases with unreacted fuel and oxygen residues are burned in the combustion chamber and then fed to heat the reformer. Exhaust gases from the cathode heat the fuel through a heat exchanger. Hot air from the reformer is used to heat the air, to heat the steam, and is directed to heat the fuel and air of the gas microturbine. The hot exhaust gases from the gas turbine heat the water and convert it into steam, part of which is used to reform the fuel. The remaining part of the steam can be supplied for technological purposes, or it can be condensed and stored in a tank of distilled water for technological needs. The hybrid system is equipped with a calciner – a unit for separating and fixing CO2 by the adsorption method. Calciner loaded with a mixture of calcium oxide with natural zeolite.

Technical and economic parameters of hybrid PP are presented in Table 2.

Table 2. Technical parameters and average cost of hybrid PP equipment.

Name	Parameters/completeness	Price
Pre-reformer	Vessel volume - 300 ml Operating pressure – 10 bar Operating temperature – 515°C Dual zone ceramic heater	\$40 000
Micro GT	Gas turbine Compressor electric generator N_{nom} = 28 kW $T_{ex.gases}$ = 295°C Operating voltage range 380-480 V Automatic control system Fuel consumption 12 m^3/h	\$80 500
30 kW AC/DC converter	Built-in fans ensure efficient heat dissipation. DC 320…460 V or 450…800 V AC 3 × 400 V or 3 × 480 V Output 10 standard voltages ranging from 24 to 800 V. The operating temperature range is -20 to +75°C.	\$16 000
SOFC	Number of cells – 1600 Fuel consumption 7.2 m^3/h fuel: methane, hydrogen air: nitrogen, oxygen rated power 30 kW operating pressure P = 1.13 bar water pump for evaporative reformer –1.17 bar electric furnace for reformer system – 400°C stack module 10×10 cm Cell voltage 0.7 V SOFC automatic control system SOFC temperature – 750°C Anode/reformer inlet temperature – 700°C Cathode inlet temperature – 600°C Anode outlet temperature – 800°C	\$193 700
Steam generator pump	Water pressure at the outlet of the pump – 2.3 bar	\$250
Steam generator	Steam pressure – 0.7 bar Steam productivity – 0.1 t/h Heating surface – 2.7m2 Fuel consumption – 1-5 m3/h Steam temperature – 130°C Efficiency – 90%	\$4 000
Distilled water storage tank	Volume – 1 m3 Quantity – 3 pcs.	\$700
Hydrogen cylinders	Volume – 100 l Operating pressure – 196 bar Reducer – 0 to 130 bar Consumption – 80 m^3	\$5 000
Methane cylinders	Volume – 100 l Operating pressure – 196 bar	\$3 000

	Reducer – 0 to 130 bar Consumption – 80 m^3	
Pre-reformer	Vessel volume – 300 ml Operating pressure – 40 bar Operating temperature – 515°C Dual zone ceramic heater	\$40 000
Calciner unit	CO_2 capture – 90% Gas temperature – 40°C Number of stages – 10 Outlet temperature –60°C Adsorbent – The amount of adsorbent –	\$1 000
Compressor for calciner unit	Pressure range – 1 – 15 bar	\$2 000
Air heater	Inlet air pressure – 3 bar Inlet temperature – 20 °C Outlet temperature – 285°C Regeneration rate – 70%	\$6 700
Fuel heater	Inlet fuel pressure – 1-3 bar Inlet temperature – 20 °C Outlet temperature – 275°C Regeneration rate – 70% Quantity – 3 pcs.	\$7 000
Automatic control system	The system includes a set of hardware and software to ensure automatic control and maintenance of the set parameters of the installation. There is a possibility of remote monitoring and data collection.	\$15 000

Table 3 presents the efficiency parameters of the SOFC-GT hybrid system operating on methane as a mini-CHPP using formulas 1 - 10 and data from Table 2.

Table 3. Technical parameters of PP as part of a methane hybrid system.

Performance indicators	**Values**
SOFC EE	58.2
SOFC-GT EE	71.3
Overall efficiency (thermal and electrical), %	99
Fuel disposal	0.8
Air disposal	1.4
The ratio of steam to carbon	2
Fuel recirculation ratio, %	75
Fuel consumption per 1 W of power, g/W	0.05
System cost per 1 kW of power, \$/kW	6900
Decarbonization degree of gaseous emissions, %	90

 https://doi.org/10.1051/e3sconf/202340401004

The amount of steam / distilled water produced, m^3	65.5

The calculated indicators correspond to the literature data. So when the SOFC is operating on methane, the EE of the fuel cell is slightly less than 60%. As is known, in the case of using pure hydrogen, the EE of the SOFC is close to 65%. By utilizing the exhaust gases in the hybrid system, the EE is increased by more than 70%. At the same time, due to the operation in the mini-CHPP mode, it is possible to achieve an overall efficiency of more than 90%.

The amount of electrochemically converted fuel in the SOFC is 80%. Air with a known oxygen concentration (~21%) is supplied to the cathode side. Thus, with a known SOFC current, the air utilization factor is 1.4.

The ratio of steam to methane is 2 and corresponds to the literature data range (1.7-2.5), at which carbon deposition does not occur.

Fuel recycling achieves two goals. First, SOFC operation is usually limited to fuel utilization in the range of 60–80% to avoid fuel depletion. If part of the fuel is reused, this increases the percentage of utilization and EE. On the other hand, a high recirculation ratio leads to a dilution of the fuel gas, which causes a drop in the Nernst voltage and system output. In this case, a recirculation ratio of 75% is used.

According to commercial offers from equipment manufacturers, the average cost of a hybrid system up to 100 kW of power is $6,900/kW. For hybrid systems in the megawatt power class, the cost varies between $3,000 and $4,500/kW.

The proposed technological scheme of the hybrid system is equipped with an additional installation for capturing carbon dioxide by the adsorption method. The adsorbent based on quicklime and zeolite has high absorption properties, low cost and environmental characteristics compared to amines. However, for systems of a higher capacity class, the adsorption method of capturing carbon dioxide may show low efficiency.

Exhaust gases from a gas microturbine can be recuperated to produce steam or distilled water if the system operates in mini-CHPP mode. According to calculations, the amount of steam is 65.5 m^3/h. Part of the steam is proposed to be used for external reforming of fuels such as complex hydrocarbons, biofuels or natural gas. In the case of a hybrid system operating on methane or hydrogen, the process occurs using internal reforming in SOFC.

4 Conclusion

The article presents a technological scheme and a feasibility study of a hybrid PP, including a gas microturbine with heat recovery, a high-temperature fuel cell and a carbon dioxide capture system. The hybrid PP is a carbon-free mini-thermal PP with integrated generation of electricity, steam and hot water with an overall efficiency of more than 90%.

For social facilities and small industrial enterprises, this technology can be an effective solution for generating electric power not exceeding 100 kW. The hybrid system can provide consumers with heat and electricity, as well as industrial enterprises with high-tech heat and steam when operating in the mini-CHPP mode.

The hybrid system shows high environmental friendliness and EE. It should be noted that such parameters are the highest among all types of PP.

The main problems of large-scale implementation still remain the high cost of SOFC as part of a hybrid system and the complexity of design with a large number of interconnections.

The results were obtained with the financial support of Ministry of Science and Higher Education of the Russian Federation "Study of processes in a fuel cell-gas turbine hybrid power plant" (project code: FZSW-2022-0001).
This work has been accomplished with financial support by Grant No BG05M2OP001-1.002-0011-C02 financed by the Science and Education for Smart Growth Operational Program (2014-2020) and co-financed by the European Union through the European Structural and Investment funds.

References

1. A. Buonomano, F. Calise, M. D. d'Accadia, A. Palombo, Hybrid solid oxide fuel cells–gas turbine systems for combined heat and power: A review. Appl. Energy, **156**, pp. 32-85 (2015).
2. P. A. Duong, B. Ryu, J. Jung, H. Kang, Thermal Evaluation of a Novel Integrated System Based on Solid Oxide Fuel Cells and Combined Heat and Power Production Using Ammonia as Fuel. Appl. Sciences, **12** (12), 6287 (2022).
3. F. Qureshi, M. Yusuf, H. Kamyab, et al, Latest eco-friendly avenues on hydrogen production towards a circular bioeconomy: Currents challenges, innovative insights, and future perspectives. Renewable Sustainable Energy Rev. **168** (2022).
4. S. Z. Golkhatmi, M. I. Asghar, P. D. Lund, A review on solid oxide fuel cell durability: Latest progress, mechanisms, and study tools. Renewable Sustainable Energy Rev. **161** 112339 (2022).
5. A. A. Filimonova, A. A. Chichirov, N. D. Chichirova, A. V. Pechenkin, Overview of design diagrams for solid oxide fuel cell and gas turbine hybrid systems for combined heat and power generation. J. Sib. Fed. Univ. Eng. & Technol. **15**(7) pp. 812–834 (2022).
6. I. K. Iliev, A. A. Filimonova, A. A. Chichirov, N. D. Chichirova, A. V. Pechenkin, A. S.Vinogradov, Theoretical and Experimental Studies of Combined Heat and Power Systems with SOFCs. Energies **16** 1898 (2023).
7. C. Viviana, M. Genovese, P. Fragiacomo, Comprehensive Review on Fuel Cell Technology for Stationary Applications as Sustainable and Efficient Poly-Generation Energy Systems. Energies **14** 4963. (2021).
8. F. Kulor, E. D. Markus, K. Kanzumba, Design and control challenges of hybrid, dual nozzle gas turbine power generating plant: A critical review Energy Rep. **7** pp. 324-335 (2021).
9. E. Facchinetti, D. Favrat and F. Marechal, Design and Optimization of an Innovative Solid Oxide Fuel Cell–Gas Turbine Hybrid Cycle for Small Scale Distributed Generation c, **14** pp. 595-606 (2014).
10. E. M. Leal, L. A. Bortolaia, A. M. Leal Junior, Technical analysis of a hybrid solid oxide fuel cell/gas turbine cycle Energy Convers. Manag. **202** 112195 (2019).
11. H. Chehrmonavari, A. Kakaee, S. Ehsan Hosseini, et al, Hybridizing solid oxide fuel cells with internal combustion engines for power and propulsion systems: A review. Renewable Sustainable Energy Rev. **171** (2023).
12. S. Giarola, O. Forte, A. Lanzini, M. Gandiglio, M. Santarelli, Techno-economic assessment of biogas-fed solid oxide fuel cell combined heat and power system at industrial scale, Appl. Energy **211** pp. 689-704 (2018).
13. Y. Cao, M, Zoghi, H. Habibi, A. Raise, Waste heat recovery of a combined solid oxide fuel cell - gas turbine system for multi-generation purposes. Appl. Therm. Eng. **198** 117463 (2021).

14. Q. Xu, Z. Guo, L. Xia, et al, A comprehensive review of solid oxide fuel cells operating on various promising alternative fuels. Energy Conversion and Management **253** 115175 (2022).
15. M. A. Abdelkareem, W. H. Tanveer, E. T. Sayed, et al. On the technical challenges affecting the performance of direct internal reforming biogas solid oxide fuel cells. Renewable Sustainable Energy Rev. **101** pp. 361-375 (2019).

Energy cluster analysis based on consumption data in different weather condition

Dimitrios Th. Kazolis[1*], *Panagiotis* G. Kogias[1], and *Nikolaos* I. Roumeliotis[2]

[1]International Hellenic University, Department of Physics, Agios Loukas, 65403, Kavala, Greece
[2]HENDO Public's Electricity Company, Kavala, Greece

Abstract. The main aim of this effort is the discovery of knowledge from data, concerning consumption of electric energy, during the year 2022, based on unattended learning methods. These data were collected from the Public Electricity Company of Kavala and the methods used are, at first the Factor analysis and second the K-means clustering algorithm. The overhead methodologies are realized by the use of Statistica Data Miner software.

1 Introduction

The purpose of all Electricity Companies is to provide stable and sufficient energy to all of their customers at any time of the year. A decisive factor in achieving this is not only the production of energy, but the prediction of future consumption and the identification of the factors that affect it. This has as result to constructive energy modeling and modifications that are essential, since nowadays; there is an increase in the energy needs of man. For this reason, there has been a growing interest in use of various techniques for extraction of knowledge from databases related to energy consumption. [1, 2, 3].

The pursuit of this knowledge is, to reveal and create conclusions, which are useful for the energy consumption process [4], [5]. Thus, there are articles aiming to the extraction of knowledge, for example in the energy consumption of buildings [6], for the characterization of electricity customers [7] or for the Analysis to Power Grid System [8]. The production of the electricity from renewable energy sources as wind farms and photovoltaic systems is also appropriate to be study [9, 10].

Generally, many data mining techniques have been implemented in the development of predicting and solving the various problems of energy consumption. [11, 12].

As a continuation of previous research, the current effort aims to contribute to the development of the specific scientific field. So, the rationale behind this work is, to exploit information that may arise from large databases.

This procedure applies in practice all key stages of the extracting knowledge process from large databases [13, 14].

Thus, a main file was created, which had 17521 lines and 20 columns. Initially, these data were changed to a numeric form. Then, they fed the process of factor analysis. This resulted in three factors and after that, the total sample led the clustering algorithm which divided it

* Corresponding author: author@email.org

into 60 groups. Those groups had a logical relevance in their categorization and proved the validity of the procedures used.

As a conclusion this article proves that, the used methods, can cooperate together perfectly in constructing understandable relationships, during the process of knowledge discovery.

2 Data and methods

2.1 Data selection

Initially, data from eight energy distribution points in the city of Kavala were recorded, with a half an hour frequency, for the entire year 2022. So, an initial table was created with 17521 lines. Then, this table combined with more variables such as, the month, the time, the days of the week, the minimum and the maximum temperature of the day, the average temperature of the month, the average humidity, the wind, the rain height, the average pressure, the sunny and rainy days and the sun intensity. [15]

As a result of the above, a final table with a total of 17521 lines and 32 columns was constructed.

Although many formats could be imported directly into Statistica, it was considered appropriate to change the format of data as many of them had a text formatting. So they were converted into Byte and Numeric type, for the better operation of the clustering algorithm.

2.2 Factor analysis

As mentioned, the result of the above actions was to create a table of 17521 lines and 32 columns. This makes a total of 543151 entries, a fact which creates great difficulty in the identification of relationships and in addition, has a negative effect on the clustering algorithm to be used. So, a multivariate exploratory technique as factor analysis was used in order to reduce the number of variables available [14]. The default minimum eigenvalues was left to be 1, and by the use of the Scree-plot and checking the Kaiser criterion, the number of the chosen factors was three. This is shown in Figure 1.

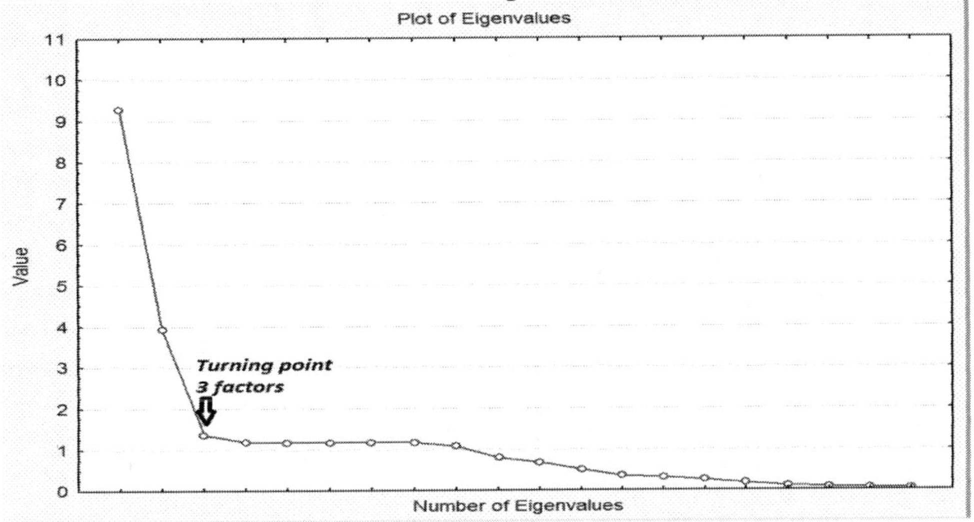

Fig. 1. Scree plot.

Furthermore a part of the correlation Matrix is depicted in the following Figure 2.

	1 TIME	2 HUM%AVG	3 hPa AVG PRESSURE	4 SUN ENERGY	TEMP. daily MAX	TEMP. daily Min	7 SATURDAY	8 SUNDAY	9 MONDAY
TIME	1,00000	-0,00000	-0,00000	-0,00000	0,00018	0,00028	-0,00000	-0,00000	-0,00000
HUM%AV	-0,00000	1,00000	0,72556	-0,83503	-0,83724	-0,80835	0,01090	-0,00390	-0,00126
hPa AVG	-0,00000	0,72556	1,00000	-0,58646	-0,71874	-0,72808	0,00802	0,00795	0,00083
SUN ENE	-0,00000	-0,83503	-0,58646	1,00000	0,69379	0,66322	-0,02057	-0,00015	-0,00184
TEMP. da	0,00018	-0,83724	-0,71874	0,69379	1,00000	0,93559	0,00403	0,01149	-0,01978
TEMP. da	0,00028	-0,80835	-0,72808	0,66322	0,93559	1,00000	0,01474	0,04619	0,00620
SATURD/	-0,00000	0,01090	0,00802	-0,02057	0,00403	0,01474	1,00000	-0,16799	-0,16799
SUNDAY	-0,00000	-0,00390	0,00795	-0,00015	0,01149	0,04619	-0,16799	1,00000	-0,16613
MONDAY	-0,00000	-0,00126	0,00083	-0,00184	-0,01978	0,00620	-0,16799	-0,16613	1,00000
TUESDA\	0,00000	0,00059	-0,00051	0,00939	-0,02433	-0,01011	-0,16799	-0,16613	-0,16613
WEDNES	0,00000	-0,00429	-0,00783	0,00939	-0,01785	-0,01484	-0,16799	-0,16613	-0,16613
THURSD/	0,00000	0,00283	0,00526	0,00367	0,01736	-0,02221	-0,16799	-0,16613	-0,16613
FRIDAY	0,00000	-0,00495	-0,01379	0,00028	0,02905	-0,02010	-0,16799	-0,16613	-0,16613
Καβάλα Σ	0,39908	0,06106	0,09774	-0,10936	-0,26231	-0,25642	-0,04758	-0,13750	0,01552
Καβάλα Σ	0,40249	0,34619	0,34258	-0,39371	-0,56199	-0,55348	-0,00846	-0,07156	0,00083
Καβάλα Σ	0,30604	-0,17564	-0,16623	0,13345	0,22656	0,21345	-0,00655	-0,15247	-0,01374
Καβάλα Σ	0,35881	-0,23160	-0,16167	0,10313	-0,02841	-0,03234	0,00903	-0,03576	-0,00661
Καβάλα Σ	0,41060	0,43312	0,39782	-0,47664	-0,62385	-0,59080	0,00833	0,01244	-0,00165
Καβάλα Σ	0,28839	0,00933	0,06760	-0,20646	-0,28596	-0,28592	0,00748	-0,00707	-0,00366
Καβάλα Σ	0,41278	0,02801	0,03771	-0,28354	-0,07996	-0,02306	-0,02143	-0,06156	0,01946
Καβάλα Σ	0,34397	0,27067	0,26237	-0,35354	-0,47160	-0,45546	-0,00551	-0,14299	0,00705
Καβάλα Σ	0,42451	0,11825	0,13028	-0,33581	-0,25192	-0,24855	0,01521	-0,06735	-0,00958
Καβάλα Σ	0,41774	0,41482	0,39029	-0,45179	-0,59244	-0,56657	0,00977	-0,01176	0,00068
Καβάλα Σ	0,41883	0,00059	0,04117	-0,11780	-0,21467	-0,20165	-0,03933	-0,15952	0,00841
Καβάλα Σ	0,40282	0,19696	0,20714	-0,36511	-0,39577	-0,35731	-0,01745	-0,07070	0,00991
Means	0,48958	69,22603	1017,43205	570,73973	20,60097	10,77123	0,14521	0,14247	0,14247
Std.Dev.	0,28862	5,93911	4,07379	184,91585	8,01529	7,44957	0,35232	0,34954	0,34954
No.Cases	17520,00000								
Matrix	1,00000								

Fig. 2. The correlation matrix.

For a factor rotation method, the Varimax raw method was used, and in the final stage of the process the loads of the factors were created. Those characterize the point that every factor interprets every variable. If a load is greater than 0.3 then it can be considered significant, but the larger it is, the more the factor interpret the corresponding variable.

So, the interpretation of the first factor correlates the time with the consumptions. As it can be observed here, there is an abnormality at the energy distribution point P230.

This is because initially, the distribution point P230 was used to distribute electric energy to the industrial and not the residential area of the city of Kavala.

The second factor interprets the change of temperature and moreover, the effect of the sun on the electricity consumptions.

Finally, the third factor interprets the variables connected with the days of the week. Here, a decline can be observed during Sundays and on the other hand, during Thursdays and Fridays the consumptions are increasing.

After the factor analysis, the 32 columns of the original file were replaced by 8 contributing to the reduction of data.

The above file is now ready to be led to the process of data mining by the use of generalized cluster analysis.

2.3 Cluster process

The main purpose of this step is to find clusters with common features, which can be interpreted by a human.

For this purpose, the Generalized k-Means Cluster Analysis is used, as, it can easily handle categorical variables. In these, the maximum frequency category grows into the center of the corresponding cluster, and all distances can be zero and one.

In this process, before the algorithm is applied, the number of clusters and the number of iterations is defined by the user. Here, the number of clusters is set to be 60 and the number of repetitions, defining group centers, is chosen to be 600. The selection of this big number was made just in order to avoid interrupting the process before it is properly completed.

Moreover, for the initial cluster centers the k observations were selected randomly and the Manhattan method is defined. This method determines the distances of non-uniformity and is the average of the modification among the elements. This is done in order to prevent the analysis from being influenced from variables with different gradient.

After all the above manipulations, a final record was created. Here, every specific hour of every month was grouped into a cluster and also was presented its distance from the cluster's center. The histogram in Figure 3 is showing the number of members of each cluster.

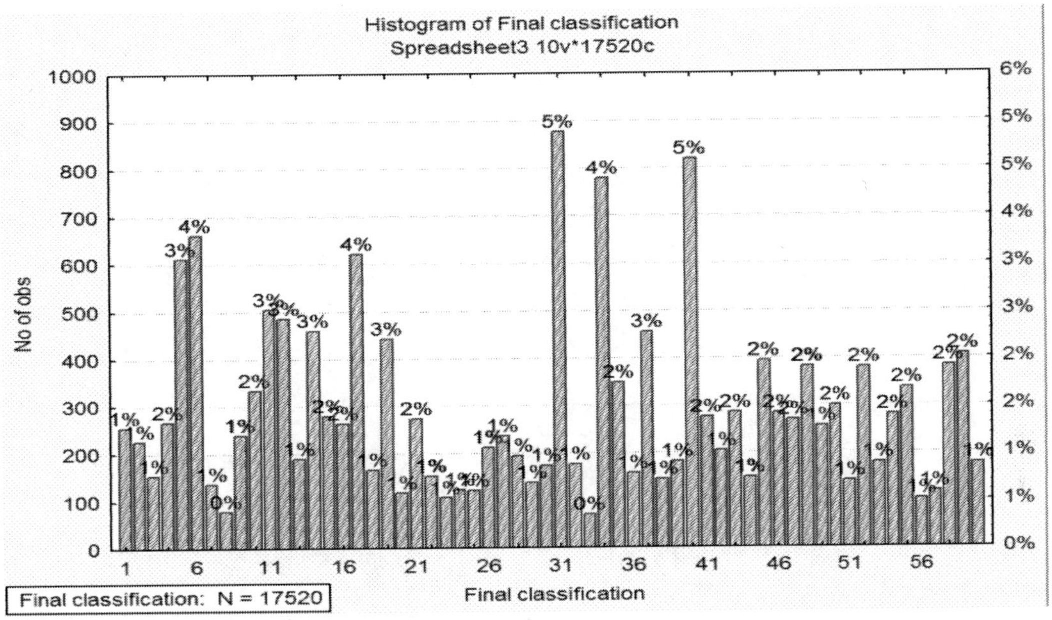

Fig. 3. Histogram of final classification.

At the present the Cluster Analysis process has been finalized. As a continuation, the examination of the similarity of each clusters' member characteristics must validate the knowledge discovery effort.

However, it should be emphasized that one of the characteristics of unsupervised learning and the K-Means algorithm is that it gives similar but not exactly the same results whenever the process occurs from the opening. And this is because the selection of the initial K centers is random.

3 Validity control

As mentioned above, the number of clusters was set to be 60, a number which can help with the interpretation of results. So, in the specific execution, some of the clusters will be examined in order to justify the reliability of this knowledge discovery process.

To begin with, cluster 2, is selected. As it can be seen in the following Figure 4, this cluster includes days of month May except Sundays. This group consists of high energy demand hours, from 8-9 o'clock in the morning until late in the evening.

	1 WIND	RAIN HEIGHT	3 SUNNY days	4 RAIN DAYS	5 tempAVG	6 FACTOR1	7 FACTOR2	8 FACTOR3	Final	Distance to
28/5/2022 9:30:00 πμ	4,6	30,2	9,7	8,9	18,9	-0,51700	-0,75651	0,00997	2,000	0,106
28/5/2022 10:00:00 πμ	4,6	30,2	9,7	8,9	18,9	-0,47415	-0,74594	0,04301	2,000	0,091
28/5/2022 10:30:00 πμ	4,6	30,2	9,7	8,9	18,9	-0,45551	-0,75260	0,04853	2,000	0,088
28/5/2022 11:00:00 πμ	4,6	30,2	9,7	8,9	18,9	-0,45191	-0,72873	0,07773	2,000	0,085
28/5/2022 11:30:00 πμ	4,6	30,2	9,7	8,9	18,9	-0,44569	-0,72378	0,05629	2,000	0,089
28/5/2022 12:00:00 μμ	4,6	30,2	9,7	8,9	18,9	-0,35702	-0,73870	0,10572	2,000	0,063
28/5/2022 12:30:00 μμ	4,6	30,2	9,7	8,9	18,9	-0,28901	-0,76721	0,16382	2,000	0,048
28/5/2022 1:00:00 μμ	4,6	30,2	9,7	8,9	18,9	-0,22323	-0,74233	0,16454	2,000	0,054
28/5/2022 1:30:00 μμ	4,6	30,2	9,7	8,9	18,9	-0,11941	-0,77295	0,16967	2,000	0,073
28/5/2022 2:00:00 μμ	4,6	30,2	9,7	8,9	18,9	-0,21023	-0,73587	0,29114	2,000	0,032
28/5/2022 2:30:00 μμ	4,6	30,2	9,7	8,9	18,9	-0,21145	-0,76247	0,24468	2,000	0,042
28/5/2022 3:00:00 μμ	4,6	30,2	9,7	8,9	18,9	-0,35900	-0,72228	0,16738	2,000	0,056
28/5/2022 3:30:00 μμ	4,6	30,2	9,7	8,9	18,9	-0,44293	-0,73988	0,09619	2,000	0,077
28/5/2022 4:00:00 μμ	4,6	30,2	9,7	8,9	18,9	-0,46636	-0,74250	0,01156	2,000	0,096
28/5/2022 4:30:00 μμ	4,6	30,2	9,7	8,9	18,9	-0,45517	-0,75226	0,02618	2,000	0,093
28/5/2022 5:00:00 μμ	4,6	30,2	9,7	8,9	18,9	-0,40830	-0,80342	-0,00296	2,000	0,104
28/5/2022 5:30:00 μμ	4,6	30,2	9,7	8,9	18,9	-0,29411	-0,85900	-0,06076	2,000	0,115
28/5/2022 6:00:00 μμ	4,6	30,2	9,7	8,9	18,9	-0,21531	-0,89229	-0,08883	2,000	0,140
28/5/2022 6:30:00 μμ	4,6	30,2	9,7	8,9	18,9	-0,13860	-0,92382	-0,05324	2,000	0,152
28/5/2022 7:00:00 μμ	4,6	30,2	9,7	8,9	18,9	-0,08366	-0,98031	-0,04777	2,000	0,173
28/5/2022 7:30:00 μμ	4,6	30,2	9,7	8,9	18,9	-0,06356	-0,95647	-0,05568	2,000	0,172
28/5/2022 8:00:00 μμ	4,6	30,2	9,7	8,9	18,9	-0,03078	-0,99162	-0,08723	2,000	0,191
28/5/2022 8:30:00 μμ	4,6	30,2	9,7	8,9	18,9	-0,00084	-1,00064	-0,07864	2,000	0,196
28/5/2022 9:00:00 μμ	4,6	30,2	9,7	8,9	18,9	0,19081	-1,04073	-0,13786	2,000	0,246
28/5/2022 9:30:00 μμ	4,6	30,2	9,7	8,9	18,9	0,19698	-1,04467	-0,10027	2,000	0,240
28/5/2022 10:00:00 μμ	4,6	30,2	9,7	8,9	18,9	0,05796	-1,00821	-0,10198	2,000	0,211
28/5/2022 10:30:00 μμ	4,6	30,2	9,7	8,9	18,9	-0,04139	-0,98508	-0,12063	2,000	0,195
28/5/2022 11:00:00 μμ	4,6	30,2	9,7	8,9	18,9	-0,15889	-0,96055	-0,14956	2,000	0,177
28/5/2022 11:30:00 μμ	4,6	30,2	9,7	8,9	18,9	-0,20214	-0,94340	-0,11363	2,000	0,160
30/5/2022 9:30:00 πμ	4,6	30,2	9,7	8,9	18,9	-0,34992	-0,70323	0,03702	2,000	0,084
30/5/2022 10:00:00 πμ	4,6	30,2	9,7	8,9	18,9	-0,34430	-0,69055	0,03332	2,000	0,087
30/5/2022 10:30:00 πμ	4,6	30,2	9,7	8,9	18,9	-0,32138	-0,68265	0,05895	2,000	0,081
30/5/2022 11:00:00 πμ	4,6	30,2	9,7	8,9	18,9	-0,34457	-0,66997	0,12390	2,000	0,074
30/5/2022 11:30:00 πμ	4,6	30,2	9,7	8,9	18,9	-0,31754	-0,66027	0,18038	2,000	0,062
30/5/2022 12:00:00 μμ	4,6	30,2	9,7	8,9	18,9	-0,29646	-0,66453	0,13917	2,000	0,067
30/5/2022 12:30:00 μμ	4,6	30,2	9,7	8,9	18,9	-0,20194	-0,68019	0,23769	2,000	0,058
30/5/2022 1:00:00 μμ	4,6	30,2	9,7	8,9	18,9	-0,13141	-0,67340	0,21432	2,000	0,074
30/5/2022 1:30:00 μμ	4,6	30,2	9,7	8,9	18,9	-0,12187	-0,67882	0,19889	2,000	0,077
30/5/2022 2:00:00 μμ	4,6	30,2	9,7	8,9	18,9	-0,20500	-0,63681	0,14924	2,000	0,085
30/5/2022 2:30:00 μμ	4,6	30,2	9,7	8,9	18,9	-0,28260	-0,65761	0,15217	2,000	0,068
30/5/2022 3:00:00 μμ	4,6	30,2	9,7	8,9	18,9	-0,40428	-0,63107	0,07012	2,000	0,103
30/5/2022 3:30:00 μμ	4,6	30,2	9,7	8,9	18,9	-0,48992	-0,58856	0,02673	2,000	0,135
30/5/2022 5:30:00 μμ	4,6	30,2	9,7	8,9	18,9	-0,43845	-0,65379	-0,05033	2,000	0,127
30/5/2022 6:00:00 μμ	4,6	30,2	9,7	8,9	18,9	-0,29318	-0,68454	0,00418	2,000	0,089
30/5/2022 6:30:00 μμ	4,6	30,2	9,7	8,9	18,9	-0,25935	-0,72535	-0,00309	2,000	0,086

Fig. 4. The number 2 cluster.

Next the cluster 38, in Figure 5, was selected. Here are days of February but the group consists of low energy demand hours, from midnight up to early in the morning.

	1 WIND	RAIN HEIGHT	3 SUNNY days	4 RAIN DAYS	5 tempAVG	6 FACTOR1	7 FACTOR2	8 FACTOR3	Final	Distance to
7/2/2022 5:00:00 πμ	4,8	49,3	7,4	8,9	6,3	-0,50202	0,78273	-1,38959	38,000	0,046
7/2/2022 5:30:00 πμ	4,8	49,3	7,4	8,9	6,3	-0,45615	0,78970	-1,38789	38,000	0,037
7/2/2022 6:00:00 πμ	4,8	49,3	7,4	8,9	6,3	-0,29398	0,75504	-1,36189	38,000	0,056
7/2/2022 6:30:00 πμ	4,8	49,3	7,4	8,9	6,3	0,00285	0,69872	-1,26842	38,000	0,132
8/2/2022	4,8	49,3	7,4	8,9	6,3	0,08878	0,85660	-1,41379	38,000	0,080
8/2/2022 12:30:00 πμ	4,8	49,3	7,4	8,9	6,3	-0,21796	0,88014	-1,43654	38,000	0,036
8/2/2022 1:00:00 πμ	4,8	49,3	7,4	8,9	6,3	-0,36519	0,88385	-1,43514	38,000	0,016
8/2/2022 1:30:00 πμ	4,8	49,3	7,4	8,9	6,3	-0,50057	0,89806	-1,41538	38,000	0,035
8/2/2022 2:00:00 πμ	4,8	49,3	7,4	8,9	6,3	-0,57850	0,88792	-1,42301	38,000	0,043
8/2/2022 2:30:00 πμ	4,8	49,3	7,4	8,9	6,3	-0,30303	0,90677	-1,41396	38,000	0,035
8/2/2022 3:00:00 πμ	4,8	49,3	7,4	8,9	6,3	-0,33926	0,89659	-1,42709	38,000	0,025
8/2/2022 3:30:00 πμ	4,8	49,3	7,4	8,9	6,3	-0,33669	0,89317	-1,41843	38,000	0,026
8/2/2022 4:00:00 πμ	4,8	49,3	7,4	8,9	6,3	-0,41167	0,88417	-1,41132	38,000	0,020
8/2/2022 4:30:00 πμ	4,8	49,3	7,4	8,9	6,3	-0,38267	0,86240	-1,40466	38,000	0,014
8/2/2022 5:00:00 πμ	4,8	49,3	7,4	8,9	6,3	-0,29794	0,84562	-1,35927	38,000	0,034
8/2/2022 5:30:00 πμ	4,8	49,3	7,4	8,9	6,3	-0,24144	0,84040	-1,32529	38,000	0,050
8/2/2022 6:00:00 πμ	4,8	49,3	7,4	8,9	6,3	-0,01262	0,81737	-1,30186	38,000	0,094
8/2/2022 6:30:00 πμ	4,8	49,3	7,4	8,9	6,3	0,26856	0,77742	-1,32067	38,000	0,141
9/2/2022	4,8	49,3	7,4	8,9	6,3	0,35322	0,89394	-1,42632	38,000	0,126
9/2/2022 12:30:00 πμ	4,8	49,3	7,4	8,9	6,3	0,03285	0,93120	-1,44436	38,000	0,084
9/2/2022 1:00:00 πμ	4,8	49,3	7,4	8,9	6,3	-0,17476	0,92926	-1,43104	38,000	0,056
9/2/2022 1:30:00 πμ	4,8	49,3	7,4	8,9	6,3	-0,29860	0,91417	-1,47350	38,000	0,033
9/2/2022 2:00:00 πμ	4,8	49,3	7,4	8,9	6,3	-0,39460	0,92097	-1,46381	38,000	0,020
9/2/2022 2:30:00 πμ	4,8	49,3	7,4	8,9	6,3	-0,13630	0,93477	-1,47932	38,000	0,064
9/2/2022 3:00:00 πμ	4,8	49,3	7,4	8,9	6,3	-0,21928	0,96057	-1,49487	38,000	0,061
9/2/2022 3:30:00 πμ	4,8	49,3	7,4	8,9	6,3	-0,23949	0,94965	-1,46937	38,000	0,050
9/2/2022 4:00:00 πμ	4,8	49,3	7,4	8,9	6,3	-0,31225	0,94465	-1,47065	38,000	0,038
9/2/2022 4:30:00 πμ	4,8	49,3	7,4	8,9	6,3	-0,28022	0,92421	-1,42709	38,000	0,040
9/2/2022 5:00:00 πμ	4,8	49,3	7,4	8,9	6,3	-0,22494	0,91205	-1,41753	38,000	0,047
9/2/2022 5:30:00 πμ	4,8	49,3	7,4	8,9	6,3	-0,13960	0,91067	-1,41938	38,000	0,059
9/2/2022 6:00:00 πμ	4,8	49,3	7,4	8,9	6,3	0,03575	0,87401	-1,34430	38,000	0,090
9/2/2022 6:30:00 πμ	4,8	49,3	7,4	8,9	6,3	0,34303	0,82484	-1,40742	38,000	0,123
10/2/2022	4,8	49,3	7,4	8,9	6,3	0,28062	0,84315	-1,32387	38,000	0,126
10/2/2022 12:30:00 πμ	4,8	49,3	7,4	8,9	6,3	0,02106	0,88632	-1,24794	38,000	0,110
10/2/2022 1:00:00 πμ	4,8	49,3	7,4	8,9	6,3	-0,22002	0,86902	-1,29202	38,000	0,062
10/2/2022 1:30:00 πμ	4,8	49,3	7,4	8,9	6,3	-0,36378	0,87421	-1,27359	38,000	0,046
10/2/2022 2:00:00 πμ	4,8	49,3	7,4	8,9	6,3	-0,46872	0,88654	-1,23868	38,000	0,063
10/2/2022 2:30:00 πμ	4,8	49,3	7,4	8,9	6,3	-0,23409	0,93813	-1,25887	38,000	0,084
10/2/2022 3:00:00 πμ	4,8	49,3	7,4	8,9	6,3	-0,24033	0,91424	-1,23842	38,000	0,081
10/2/2022 3:30:00 πμ	4,8	49,3	7,4	8,9	6,3	-0,31720	0,91797	-1,25598	38,000	0,067
10/2/2022 4:00:00 πμ	4,8	49,3	7,4	8,9	6,3	-0,38702	0,90320	-1,28162	38,000	0,048
10/2/2022 4:30:00 πμ	4,8	49,3	7,4	8,9	6,3	-0,37206	0,89413	-1,22762	38,000	0,059
10/2/2022 5:00:00 πμ	4,8	49,3	7,4	8,9	6,3	-0,30625	0,89395	-1,17389	38,000	0,079

Fig. 5. The number 38 cluster.

The following, cluster 43 in Figure 6, has the same characteristics as the previous cluster, with low energy demand hours, from midnight up to early in the morning. The clustering algorithm differentiated them because of the month August.

13/3/2022 10:00:00 μμ	5,3	36,6	6	9,2	9	1,60736	0,80396	-0,36542	42,000	0,227
1/8/2022 1:30:00 πμ	3,9	17	18,5	3,4	25,5	0,13827	-1,48801	-0,42243	43,000	[illegible]
1/8/2022 2:00:00 πμ	3,9	17	18,5	3,4	25,5	0,01589	-1,46348	-0,45243	43,000	0,077
1/8/2022 2:30:00 πμ	3,9	17	18,5	3,4	25,5	-0,08995	-1,43382	-0,47413	43,000	0,058
1/8/2022 3:00:00 πμ	3,9	17	18,5	3,4	25,5	-0,12504	-1,42488	-0,47198	43,000	0,051
1/8/2022 3:30:00 πμ	3,9	17	18,5	3,4	25,5	-0,19126	-1,41874	-0,50573	43,000	0,046
1/8/2022 4:00:00 πμ	3,9	17	18,5	3,4	25,5	-0,25603	-1,42435	-0,48766	43,000	0,048
1/8/2022 4:30:00 πμ	3,9	17	18,5	3,4	25,5	-0,28078	-1,39287	-0,41271	43,000	0,035
1/8/2022 5:00:00 πμ	3,9	17	18,5	3,4	25,5	-0,29200	-1,39811	-0,41127	43,000	0,035
1/8/2022 5:30:00 πμ	3,9	17	18,5	3,4	25,5	-0,27217	-1,41472	-0,40602	43,000	0,032
1/8/2022 6:00:00 πμ	3,9	17	18,5	3,4	25,5	-0,33136	-1,41377	-0,37580	43,000	0,034
1/8/2022 6:30:00 πμ	3,9	17	18,5	3,4	25,5	-0,30957	-1,42595	-0,37986	43,000	0,035
1/8/2022 7:00:00 πμ	3,9	17	18,5	3,4	25,5	-0,17730	-1,46624	-0,26757	43,000	0,027
2/8/2022 1:30:00 πμ	3,9	17	18,5	3,4	25,5	0,02477	-1,43817	-0,45338	43,000	0,072
2/8/2022 2:00:00 πμ	3,9	17	18,5	3,4	25,5	-0,09174	-1,40870	-0,44797	43,000	0,047
2/8/2022 2:30:00 πμ	3,9	17	18,5	3,4	25,5	-0,18076	-1,40740	-0,47950	43,000	0,040
2/8/2022 3:00:00 πμ	3,9	17	18,5	3,4	25,5	-0,26052	-1,38433	-0,46860	43,000	0,046
2/8/2022 3:30:00 πμ	3,9	17	18,5	3,4	25,5	-0,31987	-1,36483	-0,45677	43,000	0,057
2/8/2022 4:00:00 πμ	3,9	17	18,5	3,4	25,5	-0,33872	-1,38200	-0,47121	43,000	0,058
2/8/2022 4:30:00 πμ	3,9	17	18,5	3,4	25,5	-0,36495	-1,36908	-0,42976	43,000	0,057
2/8/2022 5:00:00 πμ	3,9	17	18,5	3,4	25,5	-0,41354	-1,35795	-0,42823	43,000	0,067
2/8/2022 5:30:00 πμ	3,9	17	18,5	3,4	25,5	-0,35504	-1,37985	-0,36782	43,000	0,041
2/8/2022 6:00:00 πμ	3,9	17	18,5	3,4	25,5	-0,24881	-1,41583	-0,35199	43,000	0,018
2/8/2022 6:30:00 πμ	3,9	17	18,5	3,4	25,5	-0,34558	-1,40451	-0,32906	43,000	0,025
2/8/2022 7:00:00 πμ	3,9	17	18,5	3,4	25,5	-0,13236	-1,46954	-0,22017	43,000	0,043
3/8/2022 1:00:00 πμ	3,9	17	18,5	3,4	25,5	0,06883	-1,47227	-0,34846	43,000	0,066
3/8/2022 1:30:00 πμ	3,9	17	18,5	3,4	25,5	-0,08999	-1,42835	-0,37644	43,000	0,038
3/8/2022 2:00:00 πμ	3,9	17	18,5	3,4	25,5	-0,20577	-1,40859	-0,36634	43,000	0,014
3/8/2022 2:30:00 πμ	3,9	17	18,5	3,4	25,5	-0,27630	-1,40004	-0,38823	43,000	0,028
3/8/2022 3:00:00 πμ	3,9	17	18,5	3,4	25,5	-0,32982	-1,37373	-0,33510	43,000	0,032
3/8/2022 3:30:00 πμ	3,9	17	18,5	3,4	25,5	-0,35472	-1,38547	-0,35189	43,000	0,036
3/8/2022 4:00:00 πμ	3,9	17	18,5	3,4	25,5	-0,40696	-1,37999	-0,35607	43,000	0,046
3/8/2022 4:30:00 πμ	3,9	17	18,5	3,4	25,5	-0,41783	-1,37213	-0,33724	43,000	0,046
3/8/2022 5:00:00 πμ	3,9	17	18,5	3,4	25,5	-0,41852	-1,36815	-0,28811	43,000	0,043
3/8/2022 5:30:00 πμ	3,9	17	18,5	3,4	25,5	-0,40264	-1,38319	-0,29728	43,000	0,035
3/8/2022 6:00:00 πμ	3,9	17	18,5	3,4	25,5	-0,30367	-1,40421	-0,24094	43,000	0,026
3/8/2022 6:30:00 πμ	3,9	17	18,5	3,4	25,5	-0,38734	-1,39836	-0,23513	43,000	0,041

Fig. 6. The number 43 cluster.

A similarity can be observed in the following cluster 57, in Figure 7. As with the previous groups this consists of the low energy demand hours, from midnight up to early in the morning of November days. Furthermore, the low consumption can be observed here until midday during Sundays.

	1 WIND	2 RAIN HEIGHT	3 SUNNY days	4 RAIN DAYS	5 tempAVG	6 FACTOR1	7 FACTOR2	8 FACTOR3	Final	Distance to
7/11/2022	4,1	62,3	8,2	9,7	10,9	-1,22639	0,82991	0,46176	57,000	0,044
7/11/2022 12:30:00 πμ	4,1	62,3	8,2	9,7	10,9	-1,33937	0,83264	0,42237	57,000	0,049
7/11/2022 1:00:00 πμ	4,1	62,3	8,2	9,7	10,9	-1,45597	0,84856	0,40099	57,000	0,066
7/11/2022 1:30:00 πμ	4,1	62,3	8,2	9,7	10,9	-1,51310	0,83741	0,34244	57,000	0,089
7/11/2022 2:00:00 πμ	4,1	62,3	8,2	9,7	10,9	-1,55815	0,83789	0,38158	57,000	0,088
7/11/2022 2:30:00 πμ	4,1	62,3	8,2	9,7	10,9	-1,54741	0,84133	0,38209	57,000	0,085
7/11/2022 3:00:00 πμ	4,1	62,3	8,2	9,7	10,9	-1,57331	0,85078	0,37696	57,000	0,088
7/11/2022 3:30:00 πμ	4,1	62,3	8,2	9,7	10,9	-1,55868	0,84578	0,39097	57,000	0,084
7/11/2022 4:00:00 πμ	4,1	62,3	8,2	9,7	10,9	-1,55508	0,84967	0,36691	57,000	0,087
7/11/2022 4:30:00 πμ	4,1	62,3	8,2	9,7	10,9	-1,54913	0,84169	0,43524	57,000	0,075
7/11/2022 5:00:00 πμ	4,1	62,3	8,2	9,7	10,9	-1,52379	0,84403	0,47012	57,000	0,064
7/11/2022 5:30:00 πμ	4,1	62,3	8,2	9,7	10,9	-1,45443	0,83003	0,48516	57,000	0,054
7/11/2022 6:00:00 πμ	4,1	62,3	8,2	9,7	10,9	-1,32788	0,81367	0,58569	57,000	0,036
7/11/2022 6:30:00 πμ	4,1	62,3	8,2	9,7	10,9	-1,17976	0,79521	0,64572	57,000	0,063
7/11/2022 7:00:00 πμ	4,1	62,3	8,2	9,7	10,9	-1,11218	0,81178	0,77825	57,000	0,095
8/11/2022	4,1	62,3	8,2	9,7	10,9	-1,22275	0,88650	0,52476	57,000	0,018
8/11/2022 12:30:00 πμ	4,1	62,3	8,2	9,7	10,9	-1,34452	0,87572	0,47956	57,000	0,028
8/11/2022 1:00:00 πμ	4,1	62,3	8,2	9,7	10,9	-1,42020	0,88332	0,45987	57,000	0,041
8/11/2022 1:30:00 πμ	4,1	62,3	8,2	9,7	10,9	-1,46820	0,88060	0,45072	57,000	0,050
8/11/2022 2:00:00 πμ	4,1	62,3	8,2	9,7	10,9	-1,51068	0,88374	0,42166	57,000	0,061
8/11/2022 2:30:00 πμ	4,1	62,3	8,2	9,7	10,9	-1,52205	0,88707	0,40012	57,000	0,067
8/11/2022 3:00:00 πμ	4,1	62,3	8,2	9,7	10,9	-1,52831	0,90330	0,42841	57,000	0,058
8/11/2022 3:30:00 πμ	4,1	62,3	8,2	9,7	10,9	-1,52435	0,89300	0,41318	57,000	0,063
8/11/2022 4:00:00 πμ	4,1	62,3	8,2	9,7	10,9	-1,53431	0,89186	0,43630	57,000	0,060
8/11/2022 4:30:00 πμ	4,1	62,3	8,2	9,7	10,9	-1,49529	0,87547	0,46665	57,000	0,052
8/11/2022 5:00:00 πμ	4,1	62,3	8,2	9,7	10,9	-1,47768	0,87083	0,51628	57,000	0,041
8/11/2022 5:30:00 πμ	4,1	62,3	8,2	9,7	10,9	-1,42221	0,85596	0,52977	57,000	0,034
8/11/2022 6:00:00 πμ	4,1	62,3	8,2	9,7	10,9	-1,27863	0,84918	0,60396	57,000	0,027
8/11/2022 6:30:00 πμ	4,1	62,3	8,2	9,7	10,9	-1,14316	0,82535	0,67240	57,000	0,067
8/11/2022 7:00:00 πμ	4,1	62,3	8,2	9,7	10,9	-1,07290	0,83916	0,80905	57,000	0,101
8/11/2022 7:30:00 πμ	4,1	62,3	8,2	9,7	10,9	-0,87664	0,85613	0,93952	57,000	0,151
8/11/2022 7:00:00 μμ	4,1	62,3	8,2	9,7	10,9	-2,89183	1,08267	0,13615	57,000	0,359
9/11/2022	4,1	62,3	8,2	9,7	10,9	-1,18079	0,96271	0,47864	57,000	0,044
9/11/2022 12:30:00 πμ	4,1	62,3	8,2	9,7	10,9	-1,31996	0,97602	0,43953	57,000	0,043
9/11/2022 1:00:00 πμ	4,1	62,3	8,2	9,7	10,9	-1,41989	0,98044	0,40643	57,000	0,065
9/11/2022 1:30:00 πμ	4,1	62,3	8,2	9,7	10,9	-1,48542	0,99311	0,39423	57,000	0,080
9/11/2022 2:00:00 πμ	4,1	62,3	8,2	9,7	10,9	-1,51320	0,98600	0,38698	57,000	0,084

Fig. 7. The number 43 cluster.

Similarly the cluster 9 in Figure 8, consists of days of April. There is a low energy demand period from midnight up to 7-7:30 o'clock in the morning. Moreover during Sundays, this low energy demand period, reaches the early afternoon hours.

	1 WIND	2 RAIN HEIGHT MM	3 SUNNY days	4 RAIN DAYS	5 tempAVG	6 FACTOR1	7 FACTOR2	8 FACTOR3	9 Final	10 Distance to
3/4/2022 5:00:00 μμ	4,9	33,2	5,7	9,3	13,7	-0,50114	0,00401	-0,98947	9,000	0,102
4/4/2022	4,9	33,2	5,7	9,3	13,7	-0,63727	0,12882	-1,28384	9,000	0,072
4/4/2022 12:30:00 μμ	4,9	33,2	5,7	9,3	13,7	-0,77948	0,14903	-1,33095	9,000	0,066
4/4/2022 1:00:00 μμ	4,9	33,2	5,7	9,3	13,7	-0,95569	0,16559	-1,36304	9,000	0,079
4/4/2022 1:30:00 μμ	4,9	33,2	5,7	9,3	13,7	-1,02418	0,16822	-1,37038	9,000	0,091
4/4/2022 2:00:00 μμ	4,9	33,2	5,7	9,3	13,7	-1,07922	0,17285	-1,37164	9,000	0,100
4/4/2022 2:30:00 μμ	4,9	33,2	5,7	9,3	13,7	-0,97296	0,19274	-1,35370	9,000	0,086
4/4/2022 3:00:00 μμ	4,9	33,2	5,7	9,3	13,7	-1,02794	0,18652	-1,36624	9,000	0,095
4/4/2022 3:30:00 μμ	4,9	33,2	5,7	9,3	13,7	-1,05388	0,19486	-1,37127	9,000	0,102
4/4/2022 4:00:00 μμ	4,9	33,2	5,7	9,3	13,7	-1,08607	0,17305	-1,35530	9,000	0,098
4/4/2022 4:30:00 μμ	4,9	33,2	5,7	9,3	13,7	-1,07216	0,17063	-1,29616	9,000	0,084
4/4/2022 5:00:00 μμ	4,9	33,2	5,7	9,3	13,7	-1,05420	0,16515	-1,32357	9,000	0,085
4/4/2022 5:30:00 μμ	4,9	33,2	5,7	9,3	13,7	-0,97703	0,14436	-1,28187	9,000	0,060
4/4/2022 6:00:00 μμ	4,9	33,2	5,7	9,3	13,7	-0,84201	0,12156	-1,27264	9,000	0,038
4/4/2022 6:30:00 μμ	4,9	33,2	5,7	9,3	13,7	-0,65513	0,09451	-1,19971	9,000	0,044
4/4/2022 7:00:00 μμ	4,9	33,2	5,7	9,3	13,7	-0,51962	0,11025	-1,10077	9,000	0,069
4/4/2022 7:30:00 μμ	4,9	33,2	5,7	9,3	13,7	-0,15297	0,09012	-0,91684	9,000	0,164
5/4/2022	4,9	33,2	5,7	9,3	13,7	-0,41346	0,03607	-1,10492	9,000	0,084
5/4/2022 12:30:00 μμ	4,9	33,2	5,7	9,3	13,7	-0,60521	0,06543	-1,16221	9,000	0,041
5/4/2022 1:00:00 μμ	4,9	33,2	5,7	9,3	13,7	-0,80860	0,07540	-1,18643	9,000	0,015
5/4/2022 1:30:00 μμ	4,9	33,2	5,7	9,3	13,7	-0,90196	0,07898	-1,24627	9,000	0,026
5/4/2022 2:00:00 μμ	4,9	33,2	5,7	9,3	13,7	-0,99157	0,09802	-1,23762	9,000	0,042
5/4/2022 2:30:00 μμ	4,9	33,2	5,7	9,3	13,7	-0,88272	0,11344	-1,26182	9,000	0,035
5/4/2022 3:00:00 μμ	4,9	33,2	5,7	9,3	13,7	-0,91497	0,10991	-1,22337	9,000	0,031
5/4/2022 3:30:00 μμ	4,9	33,2	5,7	9,3	13,7	-0,94252	0,09756	-1,28692	9,000	0,045
5/4/2022 4:00:00 μμ	4,9	33,2	5,7	9,3	13,7	-0,97527	0,08472	-1,26487	9,000	0,042
5/4/2022 4:30:00 μμ	4,9	33,2	5,7	9,3	13,7	-0,93682	0,07560	-1,23505	9,000	0,026
5/4/2022 5:00:00 μμ	4,9	33,2	5,7	9,3	13,7	-0,90680	0,06959	-1,21914	9,000	0,021
5/4/2022 5:30:00 μμ	4,9	33,2	5,7	9,3	13,7	-0,83983	0,05496	-1,19213	9,000	0,016
5/4/2022 6:00:00 μμ	4,9	33,2	5,7	9,3	13,7	-0,69695	0,01628	-1,17179	9,000	0,042
5/4/2022 6:30:00 μμ	4,9	33,2	5,7	9,3	13,7	-0,51171	0,01297	-1,09142	9,000	0,078
5/4/2022 7:00:00 μμ	4,9	33,2	5,7	9,3	13,7	-0,30964	-0,01607	-1,01737	9,000	0,130
6/4/2022	4,9	33,2	5,7	9,3	13,7	-0,28134	0,06522	-1,07627	9,000	0,102
6/4/2022 12:30:00 μμ	4,9	33,2	5,7	9,3	13,7	-0,53926	0,09068	-1,17556	9,000	0,056
6/4/2022 1:00:00 μμ	4,9	33,2	5,7	9,3	13,7	-0,73428	0,10162	-1,21045	9,000	0,037
6/4/2022 1:30:00 μμ	4,9	33,2	5,7	9,3	13,7	-0,85858	0,11149	-1,23683	9,000	0,026
6/4/2022 2:00:00 μμ	4,9	33,2	5,7	9,3	13,7	-0,90884	0,10434	-1,23858	9,000	0,032
6/4/2022 2:30:00 μμ	4,9	33,2	5,7	9,3	13,7	-0,81020	0,12610	-1,22181	9,000	0,034
6/4/2022 3:00:00 μμ	4,9	33,2	5,7	9,3	13,7	-0,83790	0,11248	-1,26405	9,000	0,035
6/4/2022 3:30:00 μμ	4,9	33,2	5,7	9,3	13,7	-0,87990	0,13275	-1,24579	9,000	0,036
6/4/2022 4:00:00 μμ	4,9	33,2	5,7	9,3	13,7	-0,91262	0,12716	-1,27163	9,000	0,046
6/4/2022 4:30:00 μμ	4,9	33,2	5,7	9,3	13,7	-0,89588	0,10507	-1,23079	9,000	0,029
6/4/2022 5:00:00 μμ	4,9	33,2	5,7	9,3	13,7	-0,85117	0,08549	-1,20708	9,000	0,015
6/4/2022 5:30:00 μμ	4,9	33,2	5,7	9,3	13,7	-0,80452	0,08989	-1,18670	9,000	0,019
6/4/2022 6:00:00 μμ	4,9	33,2	5,7	9,3	13,7	-0,68918	0,05233	-1,12697	9,000	0,036

Fig. 8. The number 9 cluster.

Similar examination can be performed in all of the created clusters. In general it can be noted that members of each cluster, have many common features, while on the other hand, the clusters have several differences among them, so they can be separated by the algorithm.

In general, it can be observed that, all of the clusters contain time periods with similarities in the energy consumption.

4 Conclusions

The conclusions from all of the above procedure can be derived from each step of it.

Initially it can be concluded that, a very effective statistical method for multivariate exploration, is factor analysis. With it, the large dimensional data, which must be processed, can be transformed into a relative smaller group of variables by replacing many of them, with factors. As a result the difficulties which could be raised from the very large dimensional data can be erased and moreover, the created factors include the essential meaning of the replaced variables

It can be inferred that, in such cases, Factor Analysis is considered to be necessary, as it is a reliable and effective method.

Moreover, the above procedure has also included the K-Means clustering algorithm. This is a very active method of classification. It can easily handle categorical variables especially in cases without learning examples.

To conclude with, this paper is one more proof that these two methods can cooperate together perfectly. They are following their theoretical background in a predictable way and they can be applied for further knowledge discovery research.

So, for further research, one could investigate the same data model with the method of V-fold cross validation, or to discover the effectiveness of different data mining algorithms in this common data model.

References

1. L.G.B. Ruiz, M.C. Pegalajar, R. Arcucci, M. Molina Solana, Expert Systems with Applications, **160**, 113731 (2020)

https://doi.org/10.1051/e3sconf/202340401005

2. Zhun. Yu, F. Haghighat, B.C.M. Fung, Sustainable Cities and Society, 25, pp. 33-38 (2016)
3. Y. Zhang, J. Liu, Sensors, **22** (10), 3704 (2022)
4. V. Kamaev, A. Finogeev, A. Finogeev and S. Shevchenko, *Knowledge Discovery in the SCADA Databases Used for the Municipal Power Supply System*. In: Kravets, A., Shcherbakov, M., Kultsova, M., Iijima, T. (eds) Knowledge-Based Software Engineering. JCKBSE 2014. Communications in Computer and Information Science, vol 466. Springer, Cham. (2011)
5. J.Kogan, C. Nicholas, M.Teboulle, Grouping Multidimensional Data. Recent Advances in Clustering (Springer-Verlag, Berlin Heidelberg, 2006)
6. A. Abdelaziz, V. Santos, M. Dias, Energies, **14** (22), 7810 (2021)
7. S. Ramos, J. Soares, S. Cempranel, I. Tavares, Z. Vale, Procedia Computer Science, **186**, pp. 475-488 (2021)
8. G. Grigoraş, M. Raboaca, C. Dumitrescu, D. Manea, T. Mihaltan, V. Niculescu, B. Neagu, Sensors, **23**(4), 1895 (2023)
9. C. Genç, A. Sakalli, I. Stoyanov, T. Iliev, G. Mihaylov, I. Beloev, E3S Web of Conferences, **207**, 02013 (2020)
10. C. Genç, A. Sakalli, I. Stoyanov, T. Iliev, G. Mihaylov, I. Beloev, E3S Web of Conferences, **207**, 02014 (2020)
11. S. Singh, A. Yassine, Energies, **11**(2), 452 (2018)
12. U. Fayyad, G. Piatetsky-Shapiro, P. Smyth, AI Magazine, **17** (3), pp. 37-54 (1996)
13. W. Liu, J. Zhao, D. Wang, WIREs Data mining and Knowledge Discovery, **11**(4), e1406 (2021)
14. N. Shrestha, American Journal of Applied Mathematics and Statistics, **9** (1), pp. 4-11 (2021)
15. http://www.emy.gr/emy/el/climatology/climatology_city?perifereia=East%20Macedonia%20and%20Thrace&poli=Kavala_Chryssoupoli (accessed on 10 March, 2023)

 https://doi.org/10.1051/e3sconf/202340401006

Energy efficiency index of industry in the conditions of energy transition in Bulgaria.

Dimitrina Koeva*

Department of Electric Power Supply and Electrical Equipment, Technical University of Gabrovo, Gabrovo, 5300, Bulgaria

Abstract. Regarding the energy transformation and the Net Zero Scenario, Bulgaria's effort is to maintain a coherent energy policy with the EU-28, in accordance with its geographical and climatic conditions and the current state of the energy intensive "Industry" sector. The general analysis of the current situation in the country, synthesised in this paper, is completed by an analysis of the specific technical features of the operation of nearly 1700 induction motors in two highly energy intensive enterprises: a chemical factory and a food factory. The energy losses on an annual basis can be determined by using several parameters and magnitudes. Recommendations and key factors for energy efficient investments are also presented.

1 Introduction

European Environmental Agency (EEA) defines „energy efficiency index of industry", (ODEX). This index is a weighted average of the specific consumption index of 10 manufacturing branches; the weight being the share of each branch in the sum of the energy consumption of these branches in year t and the sum of the implied energy consumption from each underlying industrial branches in year t (based on the unit consumption of the sub-sector with a moving reference year). For manufacturing industry, the evaluation is carried out at the level of 12 branches: 7 main branches: chemicals, food (beverage and tobacco), textile (and leather), wood, machinery (and metal products), transport vehicles and other manufacturing; 3 energy intensive branches: steel, cement, pulp and paper, mining and construction; 2 residual branches: other primary metals (i.e. primary metals minus steel) and other non-metallic minerals (i.e. non-metallic mineral minus cement). On this basis, the EEA prepares an energy profile for each country, [1]. This country profile summarizes key data related to industry: its relevance with respect to economic contributions, energy and water consumption, as well as air and water emissions and waste generation. Country profiles are available for EEA 33 member countries (Bulgaria is a member of group) and EU-28, Figure 1. In 2017, for example, industry in Bulgaria was responsible for 27.69% of energy consumed, 23.99% of gross value added and 83.76% of water use! The National Integrated Plan in the field of energy and climate of Bulgaria 2021-2030 has been prepared in accordance with the requirements of Regulation (EU) 2018/1999

* Corresponding author: dkoeva@abv.bg

and formulates the goals, priorities and policies for increasing energy efficiency in the country, [2].

In this regard, national targets have been set to achieve a 27.89% reduction in primary energy consumption and a 31.67% reduction in final energy consumption by 2030, relative to the reference scenario PRIMES. In quantitative terms, the country's national goals until 2030, specified in this plan, are:

- Primary energy consumption (PEC) - 17 466 ktoe (in 2021 – 18 578.1 ktoe)
- Final energy consumption - 10 318 ktoe (in 2021 – 10 247.1 ktoe)
- Final consumption in the sector "Industry"- in 2021 –2 910.7 ktoe, which is 28.41% of the total for this year.

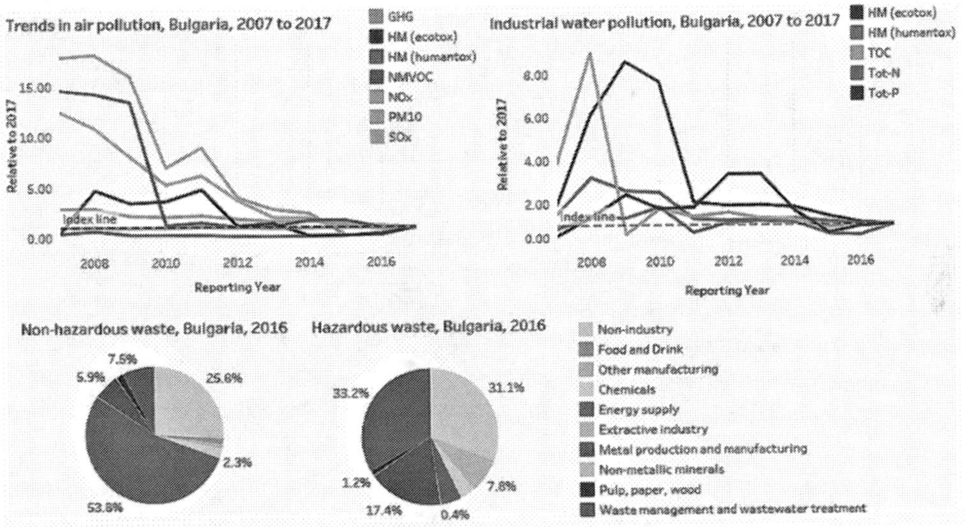

Fig.1. Energy profile of Industry sector in Bulgaria according to data of EEA.

Bulgaria's national goal in terms of energy savings is defined in the Integrated Energy and Climate Plan of the Republic of Bulgaria 2021-2030. The country uses the opportunity under Directive (EU) 2018/2002 to gradually increase the goal by reaching the defined cumulative goal of 4,357.55 ktoe in 2030. The national cumulative target is met by introducing a duty scheme for energy traders and alternative measures. All this is happening under the conditions of energy transition and activities part of the REPowerEU plan, [3]. In the Industry sector, a significant decrease of the GVA (8.6%) was observed in the crisis year 2020, as well as minimal growth in 2021. After a minimal decrease in 2020, the energy consumption grew significantly in 2021. However, the energy intensity grew in the last two years with the total growth being almost 12%. The effect of the crisis in 2020 lead to a decrease in the energy consumption, due to a decrease in the economic activity in the industry, but this decrease has been compensated by a decrease in the efficiency in production due to the crisis. In 2021, compared to 2020, the GVA grew with only 1.7%. At the same time, the energy consumption has increased by almost 10%, leading to a significant increase in the energy intensity with 8.1%.

International Renewable Energy Agency (IRENA) presents data for Bulgaria in terms of generating capacity according to the type of technology used, energy losses and electricity consumption by sector, Figure 2. [4]. The data are summarized in Table 1. When it comes to the beginning of the energy transformation period, the primary energy consumption has the main objective of limiting the use of electrical energy from non-renewable sources and increasing its efficient use in general. A major contribution to the reduction of the energy consumption is the accelerated implementation and absorption of the potential of renewable energy sources. [3, 5].

Table 1. Bulgaria energy balances, 2020.

Supply by Technology, 2020, MWh	Electricity	- 625 138.89
	Pellets	-400 277.78
	Solar Thermal	318 722.22
	Renewable Waste	487 638.89
	Other Renewable	1 297 888.89
	Geothermal	415 277.78
	Wind	1 477 138.89
	Solid Biofuels	19 114 111.11
	Solar PV	1 480 861.11
	Liquid Biofuels	2 001 027.78
	Hydropower	2 820 388.89
	Biogas	619 916.67
Total Transformation Losses, 2020, MWh	CHP	-1 908 750
	Own Use	-1 436 666.67
	Distribution	-771 333.33
	Electricity	-479 888.89
	Heat	-29 027.78
Consumptions by Sectors, 2020, MWh	Transport	2 077 194.44
	Industry	4 968 000
	Residential	12 683 472.22
	Commercial	4 403 166.67
	Other	159 944.44

The increase of the energy consumption in the industry is caused by the increase in the consumption of electricity and heat. The energy intensity decrease in the industry is due to the replacement of low-efficiency fuels (coal) with high-efficiency energies (electricity and heat). In Bulgaria, the levels of final energy intensity in the industry (FEI) are defined as high (this is a ratio of the consumed energy to GVA, created in the industry). In the last five years, there has been an increase in the GVA created in the industrial sector, under the conditions of decreasing energy consumption (Figure 3). The actual reduction of the industry's FEI is equally due to the improvements in energy efficiency and internal restructuring, concentrated on industries with lower energy intensity. Despite this downward trend, the FEI of the industry remains relatively high – about twice as high as the average for EU countries, Figure 4, [4].

 https://doi.org/10.1051/e3sconf/202340401006

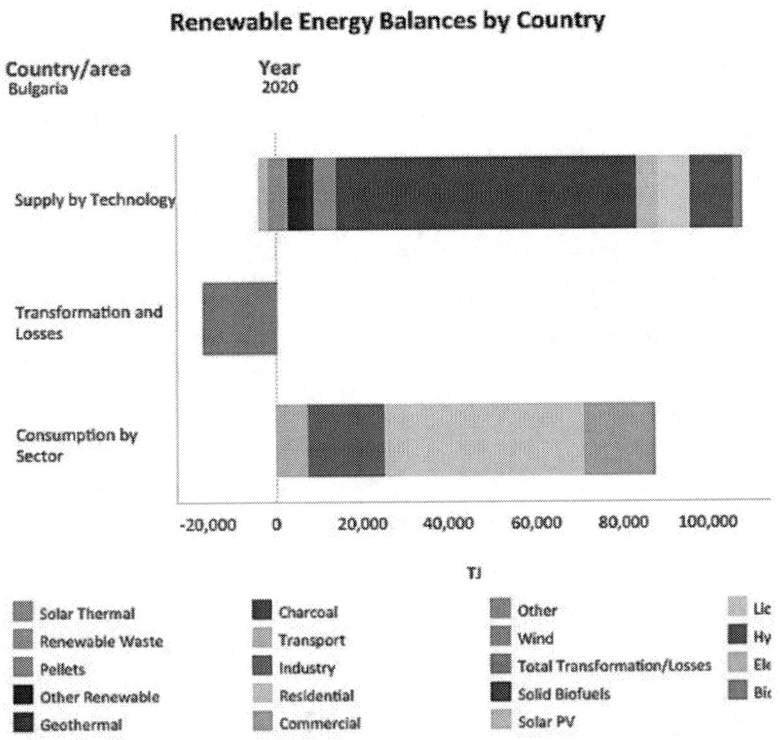

Fig. 2. Bulgaria energy balances. (https://www.irena.org, last update: 09 August 2022).

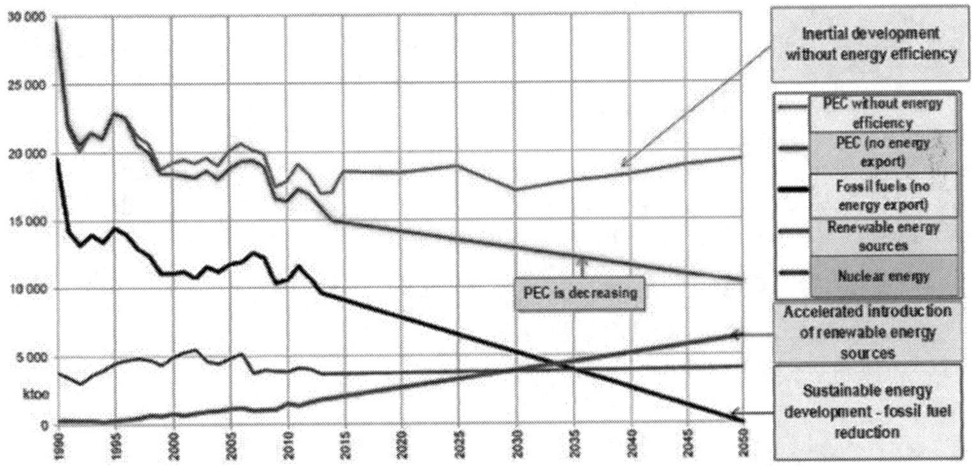

Fig. 3. PEC according to Integrated plan in the field of energy and climate of Bulgaria.

The most frequently recommended measures for increasing energy efficiency are:
- Combined production of heat and electricity (cogeneration) – 1%;
- Use of RES – 2%;
- Reduction of energy costs for buildings – 11%;
- Energy-saving lighting – 12%;
- Energy efficient transformers – 5%;
- Introduction of highly energy efficient electric drives and motors – 6%;
- Changing the used fuel – 5%;
- Increasing the energy efficiency of boiler plants - 8%;
- Improvements in thermal insulation (14%) and steam transferring systems (8%);

- Improvement of technologies (11%) and/or replacement of technological equipment (10%);
- Energy monitoring systems – 6%.

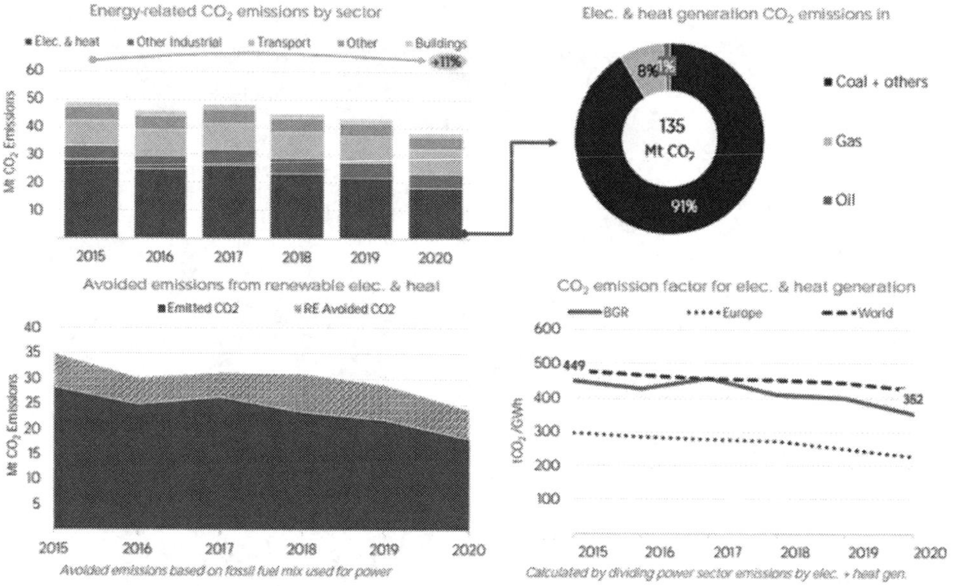

Fig. 4. Energy related CO2 emissions by sector in Bulgaria.

Energy efficiency measures in the industry do not go out of focus. On the contrary, energy efficiency management in all enterprises and industrial systems becomes mandatory. This measure has been regulated in Art. 63 of the EE Law, according to which EE management should be carried out. The main object of one of these measures is industrial systems, which annual energy consumption is over 3000 MWh. For 2022, 57 industrial systems have been inspected. Energy saving measures (ESM) for lighting installations and replacement of technological equipment have been marked as the ones that have been used the most. In 2022, the largest amount of energy was saved by applying the ESM measure on technological aggregates and equipment and by the replacement of technological equipment. According to the assessment of the effect of energy efficiency management for the past year (2022) in enterprises and industrial systems the saved energy is GWh/y 449.6; the saved carbon emissions CO2 ktCO2/y are 187.2; the investments are 130.9 million lv/y, as 78% of them are own investments.

2 Technical considerations

The EE measures that have to be taken under consideration are focused in two directions. The first is requiring large investments and optimizing electricity consumption in the long run and it includes identifying and implementing measures, related to limiting electricity consumption. With the second alternative, the industrial site maintains its current consumption, but uses all the opportunities provided by the liberalization of the energy market. The choice of specific measures is strictly individual and fully compliant with the specifics of each enterprise. The development of an effective program begins with an objective and accurate analysis of the consumption of energy resources. In order to analyze, it is necessary to provide information on the actual consumption of all capacities, included in the relevant production. In order to provide this information, comprehensive systems are being built to measure the consumption in all key locations – aggregates, transformers, substations, production lines, etc., as well as the industrial facility itself as a whole.

The second stage includes the development of the program itself, as here a very important element is the analysis of the current consumption structure. This is essential in order to predict future consumption. It is also necessary to analyze the technical condition of the existing technological equipment, as well as to build a system for measuring and controlling the consumption at all critical points. The analysis is both technical and economic, as it takes into account the profitability of the planned investments, considering the expected electricity savings. Another approach, that specifically refers to the optimization of electricity costs, involves selecting the most attractive price tariff for the payment of electricity.

The electricity consumption of the industrial users is mostly measured with three-tariff electricity meters, due to the large amounts of energy that they consume. In order to reduce the costs that the enterprise has to pay, it is necessary to precisely calculate the peak consumption of day and night electricity. The ultimate goal is to manage the load schedule so that the most energy-intensive activities are concentrated in the periods of cheapest energy. Therefore, it is possible to impose a non-standard mode of operation of the enterprise in order to reduce the costs. However, managing the load schedule by trying to load the production capacities in the hours of cheaper energy is not always technologically possible. On the other hand, switching to a non-standard mode of work is associated with higher wage costs. For these reasons, the implementation of this type of event is recommended after expert analysis has been carried out.

When talking about the optimization of electricity costs, the liberalization of the electricity market should not be overlooked, i.e. using the lowest possible electricity price. Another approach to reduce the costs of electrical energy in enterprises without limiting the consumption is to implement measures that are directed towards reaching compliance between the regulated and the actually achieved power factor. This will save the allowances that is paid every month depending on the registered deviation between the prescribed and the achieved power factor.

The second major approach to reduce the energy costs involves developing and implementing a program to limit the consumption. This could be achieved by technological optimization, loss limitation or by a very good knowledge of the technology specifics. It usually works both ways. Among the largest consumers of electrical energy in the field of industry are electric motors, as a predominant share of them have IM. It is considered that if the average load of the engine is 45% of the nominal, it is economically feasible to replace it with a model with less power. In applications where the motor load is higher than 40% and at the same time lower than 70% , the decision to replace it is made after a thorough analysis. Initially, the calculation of the real load factor depends on the operation mode of the engine - with a relatively even loading, with cyclic loading without switching off the engine between cycles, with cyclic loading with switching off the engine at the end of the cycles and with non-periodic and non-cyclic loading. After the load factor has been calculated and depending on the so-called power losses, it can be determined how appropriate the investment in a new engine is.

3 Power and energy losses in IM with industrial application

The electrical equipment of industrial units, and in particular their electrical drives, are the main consumers of electrical energy in the industry. Electric drives are included in the electrical equipment of mass-used fans, pumps, conveyors, cranes, elevators, machine tools, etc. The development of electric drives does not exclude but anticipates the development of other types of industrial drives (hydraulic, pneumatic, internal combustion engines, etc.). In some cases, combined drive systems are used, e.g. electric and hydraulic drives for electric trucks, high-lift trucks and machine tools, etc. By using an established mode of operation of the electric motors, the losses of active energy ΔW_a in them can be determined as the sum of the losses in the windings in the steel and also the mechanical losses:

$$\Delta W_a = 3.k_{fi}^2.I_{cp}^2.R_e.T_p, \quad (1)$$

where R_e for IM is the sum of the stator resistance and resistance of the rotor: $r_1 + r_2^{`}$.

Steel losses are determined by measurement only for wound-rotor IM:

$$\Delta W_{steel} = T_{act}(P_0 - 3.I_{10}^2.r_1), \quad (2)$$

where P_0 - idle power; I_{10} - idle current, T_{act} - actual working time; k_{fi} - form factor of the load schedule by current.

For all motors (except for wound-rotor induction motors), steel losses should not be separated into a different part of the energy balance. Like the mechanical losses, they also depend on the load of the electric motor, and therefore they are combined with them. The mechanical losses in the units and the electrical losses in the steel of the AC drive motor are determined by the equation:

$$\Delta W_{mech} + \Delta W_c = (P_{id} - 3.I_{id}^2.r_1).T_{act}, \quad (3)$$

where: P_{id} - no-load power of the electric motor coupled to the drive mechanism; I_{id} – idle current.

Losses in transient modes of electric motors are determined as follows:

- startup process $\Delta W_п \approx k.h_п.\frac{GDn_0^2}{2620}.10^{-6}$, [kWh];
- when stopping by switching $\Delta W_{тп} \approx 3кh_т\frac{GD^2n_0^2}{2620}.10^{-6}$, [kWh];
- in reverse braking mode $\Delta W_{рев} \approx 4k.h_p.\frac{GDn_0^2}{2620}.10^{-6}$, [kWh];
- in dynamic braking mode $\Delta W_{тп} \approx \Delta W_п$,

where: $h_п$, $h_т$, h_p - number of starts, stops, reversals for 1 hour; n_0 – speed of the electric motor at ideal idle speed; k=1+$\frac{r_1}{r_2}$ for IM; GD^2 - motor and mechanism torque.

4 Calculation and results

The subject of the study is the operation of IM in two industrial enterprises whose production is defined as energy-intensive (chemicals and food - two of the main 7 branches). The losses in the electric motors of these enterprises, which for technological reasons work from 9 to 22 hours a day and 335-348 days in the year (in three-shift operation mode) are calculated. The form factor of the load current schedule k_f is determined as follows: for IM operating from 9 to 12 hours - k_f = 0.86; operating from 12 to 15 hours - k_f=0.88; operating from 15 to 21 hours - k_f=0.9 and operating until 22 hours - k_f=0.92. In order to determine the losses in detail, we have previously defined the parameters and magnitudes of the IM that drive the industrial systems in both enterprises using ANSYS MotorCAD motor design software, [6, 7, 8]: the resistance of one phase of the stator winding r_1, rotor resistance reduced to stator winding $r_2^{`}$, idle power P_{id} and the idle current I_{id}, Table 2.

Table 2. Parameters and dimensions of the considered IM.

IM rated power, kW	**0.4**	**0.8**	**1.1÷2.2**	**4**	**7.5**	**11**	**13**	**15**	**17**	**18.5**	**22**	**75**	**up to 180**
r_1, Ω	35.1	20.13	8.39	1.7	1.37	0.52	0.36	0.402	0.318	0,261	0.22	0.123	0.071
r_2', Ω	21.55	6.3	3.97	0.942	0.57	0.41	0.245	0.196	0.158	0,153	0.14	0.01	0.088
P_{id}, W	38	73.87	109.48	275.75	243.6	402	392	435.8	362.8	618,2	754.5	927	1537
I_{id}, A	0.5	0.77	0.48	0.42	5.6	6.2	7.24	7.8	6.95	9.24	12.75	33.5	39.75

The total number of drive motors in the Chemical factory is 1047, and in the Food factory – 664 as they have been listed according to their rated power and presented in Table 3. The active energy losses in the motors in the Chemical factory are 7.65% of the total annual active electrical energy consumed, and the active energy losses in the motors from the technological production in the Food factory – 11.5%, Figure 5. Reactive energy losses in engines in the Chemical factory are 4.85% of the annual amount of reactive energy consumed, and for the Food factory - 9.03%, Figure 6. Almost twice as many engines, which have a significant share in the energy losses of the Food factory! What could be the reason for that?

Table 3. IM according to rated power and application.

IM rated power, kW	**0.4**	**0.8**	**1.1÷2.2**	**4**	**5.5**	**7.5**	**11**	**13**	**15**	**18.5**	**22**	**30**	**55**	**75**	**up to 180**
Chemical factory	434	18	216	4	134	154	23	6	6	2	10	4	19	3	14
Food factory	344	10	157	38	18	13	22	-	18	10	11	6	6	7	4

Obviously, the assessment of purely technological options for reducing the consumption is a complex process that requires a profound knowledge of the currently available technologies. Each production is characterized by the so-called theoretical minimum in energy consumption. In order to make a precise analysis of the opportunities for savings, the consumption in the process of implementing the technology and the consumption in the process of its operation must be known. The lower the energy consumption of a given technology, the greater the opportunities for energy savings. On the other hand, it is also true to say that the less energy-intensive a given technology is, the more expensive and difficult it will be to implement the measures for further reduction of the energy consumption. Usually, the specific consumption of electricity in the process of implementation of a given technology has the highest value. In the process of operation and due to a number of factors (among which is the knowledge of the technology specifics by the personnel) the specific cost decreases. Eventually, the main question which technology

 https://doi.org/10.1051/e3sconf/202340401006

to choose and whether to keep the old one by trying to increase the energy efficiency, or to initiate a complete replacement with a lower energy-intensive technology is purely economic.

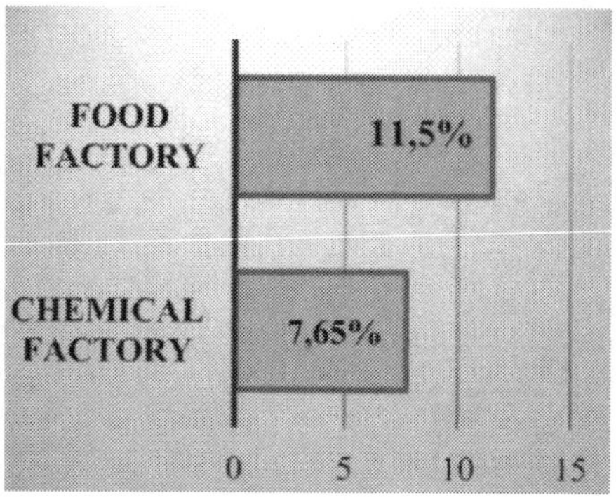

Fig. 5. Active energy losses.

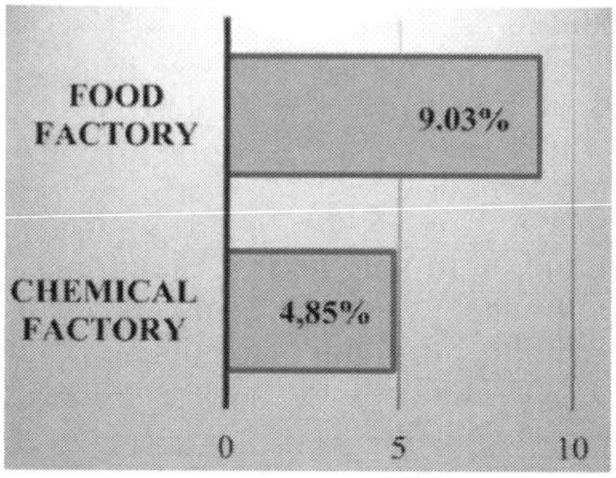

Fig. 6. Reactive energy losses.

That is one more reason to emphasize on the great importance of some key economic evaluation indicators, when it comes to investing in energy efficiency, [9]. These indicators are static and dynamic. For example: a simple redemption period (PB) is static and does not reflect the moment when the investments are made; O/M – maintenance costs, etc. The relation between them is described by the following equations:

$$B_b = S.E; \tag{4}$$

$$B = S.E - \frac{O}{M}; \tag{5}$$

$$PB = \frac{I_0}{B}, \tag{6}$$

where: B – economy of energy saving activities, [kWh/year, kWh/lv]; E- average price, [lv/kWh/year]; S - saved energy for 1 year; B_b - gross annual savings; B - net annual savings; PB - simple redemption term; I_0 - initial investment; $\frac{O}{M}$ - maintenance costs.

There are also dynamic indicators: n - asset utilization period; r- real rate of interest, %, (software that estimates energy savings consider this a real estimate, as it is adjusted for inflation).

$$r = \frac{r_n - b}{1 + b}; \tag{7}$$

$$PV_i = \frac{B_i}{(1 + r)^i}; \tag{8}$$

$$NPV = B.\frac{1 - (1 + r)^{-n}}{r} - I_0, \tag{9}$$

where: r_n - nominal interest rate, %; b – inflation %; i - year being observed; PV_i- present value of net savings, [lv]; $\frac{1}{(1+r)^i}$ - the discount factor; NPV - net present value of the investment, [lv], (if $NPV > 0$, then the investment has been successful).

The difference between the two energy-intensive enterprises in this polar example comes down to:

- responsible and practical use of the ISO: 50001 - Energy management standard;
- use of secondary energy sources;
- use of the above-mentioned key dynamic indicators for energy efficient investments;
- consideration of the fact that the investment risk assessment includes technical, economic, administrative and management indicators. A simplified assessment of the energy efficiency can be carried out when production growth is maintained. Then the energy efficiency matches the energy intensity, [6].

If the production parameters are changed (a frequently observed situation during an energy crisis and energy transition), a recalculation of the average annual, initial and expected consumed electrical energy is required. If it is necessary to determine the ecological footprint of an average economy on an annual basis, then for saved energy $EF = 1.039\ kgCO2/kWh$ and for generated energy $EF = 0.833\ kgCO2/kWh$, [10].

Industrial activity in 2021 was directly responsible for emitting 9.4 Gt of CO2, which is equal to a quarter of the global emissions (without including the indirect emissions from electricity used for industrial processes). In the Net Zero Emissions by 2050 Scenario, industrial emissions fall to about 7 Gt CO2 by 2030, despite the expected industrial production growth. Modest improvements have already been made in the energy efficiency and also in renewable energy uptake, as well as some positive policies and innovation steps [5, 10].

Every EU country has its own strategy according to the level of its economic development. In our country, a new energy strategy must be built along with the development of an industrial strategy, given the forecast for an increase in the consumption of the energy-intensive industry. But before that it is necessary to make a detailed and realistic analysis of both the supply of electricity and the consumption. An important condition is to guarantee the provision of electricity at such a price that will preserve the competitiveness of the Bulgarian industry and will promote our native production.

It is important to point out the difference between energy amplification and energy intensity. Energy amplification is when a lot of energy is consumed in general, and energy intensity is when a lot of energy is consumed per unit of product due to the characteristics of the technological processes - electrolysis, melting of metals, synthesis of chemical products, etc. Energy-intensive industries are efficient, but they require a lot of energy and therefore the needed resources form a major component of their costs.

For many years the industry in Bulgaria has been the most energy efficient sector by constantly improving its performance. All large enterprises are continuously working towards improving their energy efficiency, because it is of major importance to good competitiveness and high export potential. Moreover, energy efficiency does not always lead to the reduction in electricity consumption. This is the so-called Jevons paradox - when the energy consumption per unit of product is reduced and the energy efficiency is increased, its production can be expanded by producing more units of the final product more efficiently and eventually increasing the energy consumption in absolute units. Despite the temporary calming of the energy markets, there is still a certain level of concern regarding the further increase in gas and electricity prices, as it cannot be determined whether the period of free and unpredictable formation of these prices has finished. Another challenge is the revival of long-term electricity deliveries to industrial end-users, as the risks arising from the high proportion of spot trading over the past 18 months have become clear.

However, what is Bulgaria's plan and isn't it too late to react? Bulgaria needs to increase the investments for energy transition. Smooth decarbonisation must be combined with very good predictability of electricity and energy resources also for the industry. Of course, with an increasing share of the energy mix of wind, solar and biomass.

The implementation of this paper is thanks to the Project BG05M2OP001-1.002-0023 Competence Centre “Intelligent Mechatronic, Eco- and Energy-Saving Systems and Technologies”, Laboratory C 7-2 "Electric drive and electrical equipment - modern energy-efficient electrical components and systems with application in the industrial sector", financed by the Operational Program ‘Science and Education for Intelligent Growth’ of the European Regional Development Fund.

References

1. ETC/ATNI Report 6/2019: EEA-33 Industrial Emissions Country Profiles. Methodology report. Eionet Report - ETC/ATNI 6 (2019)
2. Integrated Energy And Climate Plan Of The Republic Of Bulgaria 2021–2030
3. REPowerEU: affordable, secure and sustainable energy for Europe: Clean Industry, (2022)
4. IRENA. Bulgaria energy profile-Energy and emission, August, 2022 (https://www.irena.org/Data/Energy-Profiles)
5. IRENA, World Energy Transitions Outlook 2022: 1.5°C Pathway, International Renewable Energy Agency, Abu Dhabi (2022)
6. International Renewable Energy Agency: Methodology-Annex II: Avoided Emission Calculator (August 2022)
7. I. P. Kopilov et al., Design of Electrical Machines, Part 1, Energoatom Publishing, Moscow **240-454** (2001)
8. R. Harwood, On the verge of an electric revolution (ANSYS Inc. CADFEM Review) **6,** 6 (2019)
9. IRENA and JRC, Benchmarking Scenario Comparisons: Key indicators for the clean energy transition, International Renewable Energy Agency, European Commission's Joint Research Centre, Abu Dhabi, Brussels (2021)
10. IEA, Direct CO2 emissions from industry in the Net Zero Scenario 2000-2030, IEA, https://www.iea.org/data-and-statistics/charts/direct-co2-emissions-from-industry-in-the-net-zero-scenario-2000-2030

https://doi.org/10.1051/e3sconf/202340401007

Methodology for energy savings assessment, at engine oil change of road vehicles

Angel Terziev[1*], *Penka* Zlateva[2], and *Martin* Ivanov[1]

[1]TU Sofia, Department of Power Engineering and Power Machines, K. Ohridski blvd. 8, 1000 Sofia, Bulgaria
[2]TU Varna, Department of Thermal Engineering, 1 Studentska Str., 9010 Varna, Bulgaria

Abstract. The presented paper reveals the possibilities of the implementation of a methodology for evaluating energy savings, at changing the engine oil of different road vehicles. The purpose of the developed methodology is to provide an opportunity to evaluate the energy consumption before and after changing the engine oil of the vehicle. The conditions, the practical procedures and the mechanism for evaluating the energy savings after changing engine oils of vehicles are defined. Changing the engine oil of the car covers the replacement of the old oil with a new one within the maximum mileage determined by the manufacturer. The results of the application of the methodology for certain categories of cars allow an easy process of monitoring and verification. And finally, a computational example is considered, proving that the methodology gives correct results for the evaluation of energy savings, after the replacement of the engine oil of the vehicle.

1 Introduction

In recent years, there has been an increase in the requirements towards one of the main goals and policies of the European Union (EU) namely, to reduce CO_2 emissions and other greenhouse gases (GHG). In this concern, the EU is moving towards a carbon-neutral economy by 2050. It also adopts both a 2030 climate and energy framework agreed with EU member states as an economy-wide reduction in greenhouse gases of 40% by 2030. Thus, shares of 32% for renewable energy and 32.5% for improving energy efficiency, are accepted. Subsequently, the European Commission also determined the emissions from transport: a 20% reduction from the levels from 2008 to 2030 and a 60% reduction compared to the levels from 1990 to 2050. Transport was considered as a significant source of emissions in 2016 as well and contributed to 27% of total EU greenhouse gas emissions, with road transport accounting for 72% of them. It follows so far that a reduction in fuel consumption in the transport sector has an impact on the total emissions of the system. In 2017, the European Commission set CO_2 standards from 2021 to 2030 for new cars and vans, and average CO_2 emissions must be 37.5% - 31% lower than in 2021 for 2030. [1-3].

* Corresponding author: aterziev@tu-sofia.bg

Technological improvements in lubricants for vehicle engines have had a significant positive effect on the reduction of greenhouse gases and CO_2 emissions in recent years [4-8]. In order to reduce emissions, the Association Technique de l'Industrie Europeenne des Lubrifiants (ATIEL) commissioned Ricardo Energy & Environment to assess historical and future CO_2 savings based on advances in engine lubricant technology through the SULTAN model. Based on the US road vehicle fleet and according to the criteria set in [9], it is estimated that as a result of engine lubrication technologies since 2005 until 2020 respectively, up to 1.2 – 3.9 $MtCO_2e/y$ avoided emissions were obtained and will continue to reduce emissions by another 0.9 – 2.7 $MtCO_2e/y$ until 2030 [9].

In order to reduce the energy savings, measures applied by EU member states have been generated, according to Article 7 of Directive 2012/27/EU on energy efficiency, related to the methods for calculating energy savings [10]. In 2018, Directive 2012/27/EU was amended by Directive 2018/2002, which sets out a general framework of measures to promote energy efficiency in the US to ensure that this energy efficiency target is met by the most at least 32.5% for 2030 [11-12]. According to the requirements of the directive, Member States need to set their own national energy efficiency targets and increase the growth of cars with e-drives [13].

In March 2021, changes were made to the Law on Energy Efficiency, which regulated the way in which the national goal for energy efficiency in Bulgaria is determined until 2030. Both obligations for energy savings and alternative measures were introduced, which ensure the achievement of the target for energy savings in final energy consumption for the period from the beginning of January 2021 to the end of December 2030 [14].

According to the authors, reference [12] includes general requirements and key principles that should be taken into account in the methodologies for calculating energy savings in EU member states.

The main goal of the presented study is to develop a methodology for evaluating the energy savings when changing the engine oil of road vehicles in Bulgaria, in accordance with the legislation of the country and the EU. The methodology defines the conditions, the practical procedures and the mechanism for evaluating the energy savings after changing engine oils of vehicles. The results of the research will be useful in finding solutions to reduce greenhouse gas emissions in line with EU climate change legislation.

2 Materials and Methods

2.1 General conditions in the development of the methodology

Considering the presented requirements of the World, European and Bulgarian legislation for the reduction of greenhouse gases and CO_2 emissions, a methodology has been developed for evaluating the energy savings when changing the engine oil of vehicles. In order the calculation procedure of the methodology to be correct, it must meet the following requirements:

The methodology defines the conditions, practical procedures and the mechanism for evaluating energy savings after changing engine oils of vehicles;

Changing the engine oil of the car covers the replacement of the old oil with a new one within the maximum mileage determined by the manufacturer;

The purpose of the methodology is to provide an opportunity to evaluate energy consumption before and after changing the engine oil of the vehicle;

The methodology defines the main steps and principles of the procedure for proving the energy savings achieved by a third party;

The results from the methodology application are proven by the New General Requirements Document for Light-Duty and Heavy-Duty ACEA Oil Sequences. This revision includes the references [15] and [16] for certain categories of cars. They enable the implementation of a simple monitoring and verification process.

The methodology allows for the possibility of estimating the energy savings after the replacement of the vehicle's engine oil to be calculated numerically.

2.2 Application characteristics of the energy service

Scope of the energy service offered

The methodology is only applicable to vehicles of categories M1, M2 and N1, which were first registered after 2007, i.e. Euro 4, Euro 5 and Euro 6.

The fuel economy is according to item 2.5 (New General Requirements Document for Light-Duty and Heavy-Duty ACEA Oil Sequences. This revision includes the references [15] and [16] and applies to SAE XW-30 oils; XW-20, or directly quoted ASEAN class and having a minimum fuel economy (petrol or diesel) greater than 2.5%.

Application sector

The methodology is applicable to the assessment of saved energy after changing the engine oil of a group of cars of a corresponding category in the field of transport.

2.3 Characteristics of the proposed methodology

The methodology is based on the application of calculation procedures that are also applied in other countries.

The calculation procedure includes several basic steps applied after processing analytical information on the number of vehicles of categories M1, M2 and N1, which are first registered after 2007, i.e. Euro 4, Euro 5 and Euro 6, and use engine oils SAE XW-30; XW-20.

The lubricating properties of energy-efficient oils and their role in maintaining better operating parameters of engines are taken into consideration. The calculation procedure is as follows:

Calculation of supplied energy with fuels (gasoline, diesel, propane-butane, methane);

Calculation of energy consumption after changing the engine oil of the road vehicle (RV);

Calculation of the amounts of final energy saved;

Calculation of the amounts of CO_2 emissions saved by the end user.

2.4 Additional information for methodology application

Terminology

T- period of action (technical life) – is the kilometers traveled after the implementation of the measure, during which it is assumed that the engine works economically, according to the action for which they were designed and lead to measurable results on energy consumption.

f_r – coefficient of reduced fuel consumption after changing the engine oil, %

The coefficient accounting for the reduced fuel consumption of the car after the engine oil change is found in accordance with item 2.5. (New General Requirements Document for Light-Duty and Heavy-Duty ACEA Oil Sequences. This revision includes the references [14] and [15] applies to SAE XW-30 oils; XW-20 or directly quoted ASEA class and having a minimum fuel economy (petrol or diesel) greater than 2.5%.

Coefficient of reduced fuel consumption after changing the engine oil is 0.975, f_i – Emission factor (t_{CO_2}/kWh) (Calculated according to Table 1 [17]);

Table 1. Reference values of the coefficient of ecological equivalent of energy/fuel.

Type of energy/fuel	f_i tCO_2/GWh
Industrial gas oil, diesel	267
Propane-butane	227
Methane	202
Gasoline	263

e_p – coefficient accounting for the losses of extraction/production and transmission of energy and energy resources ((Found according to Table 2 [17]);

Table 2. Reference values of the coefficient accounting for losses for extraction/production and transmission of energy, including fuels.

Type of energy/fuel	e_p
Industrial gas oil, diesel	1.1
Propane-butane	1.1
Methane	1.1
Gasoline	1.1

ρ_{fuel} – fuel density, (kg/m^3) (Calculated according to Table 3 [18]);

LHV_{fuel} - lower heating value of combustion of fuel, (kWh/kg) (Calculated according to 3 [18]).

Table 3. Reference values of fuel density and lower heating value of combustion of fuel (LHV), for the different types of fuels.

Type of fuel [-]	Density kg/l	LHV of combustion kWh/kg
Diesel	0.832	11.860
Gasoline	0.745	12.209
Propane-butane (LPG)	0.538	12.791
Methane (CNG)	0.554	13.100

D – average mileage of a vehicle of the respective category, (km/y) .

N – number of registered vehicles of the respective category (Euro 4, Euro 5 and Euro 6) with engine oil changed SAE XW-30; XW-20;

D_a – average annual mileage of the vehicles of the respective category, (km/y)

F_B – average fuel consumption of all vehicles of the respective category (i=1,2,3), l/100 km;

Q_B – annual fuel consumption per vehicle category before engine oil change, (l/y);

FE_B – energy delivered with the fuel for the respective vehicle category before the engine oil change, (MWh/y);

Q_A – annual fuel consumption for vehicle category after engine oil change, (l/y);

F_A– average fuel consumption of all vehicles of the respective category after changing the engine oil (i=1,2,3), l/100 km, where:

FE_A– delivered energy with the fuel for the respective category of vehicles after changing the engine oil, (MWh/y);

PES – primary energy saved, MWh/y;

2.5 Calculation of energy savings according to the methodology

The methodology takes into account the difference between the energy used before and after the introduction of new energy-efficient motor oils.

2.5.1 Calculation of the energy saved and reduced carbon emissions when changing the engine oil of a car

Energy savings

1. **Average annual mileage for all cars:**

$$D_a = N * D\ (km/y), \tag{1}$$

where:
D_a – average annual mileage of the vehicles of the respective category, (km/y)
N – number of registered vehicles of the respective category (Euro 4, Euro 5 and Euro 6) with engine oil changed SAE XW-30; XW-20;
D – average mileage of a vehicle of the respective category, (km/y) .

2. **Annual fuel consumption for the relevant vehicle category before changing the engine oil:**

$$Q_B = F_B * D_a\ (l/y), \tag{2}$$

where:
Q_B – annual fuel consumption per vehicle category before engine oil change, (l/y);
F_B – average fuel consumption of all vehicles of the respective category (i=1,2,3), l/100 km;
D_a – average annual mileage of the vehicles of the respective category, (km/y)

3. **Delivered energy with fuel before engine oil change**

$$FE_B = Q_B * \rho_{fuel} \cdot LHV_{fuel}\ (\mathrm{MWh}/y), \tag{3}$$

where:
FE_B – energy delivered with the fuel for the respective vehicle category before the engine oil change, (MWh/y);
Q_B – annual fuel consumption per vehicle category before engine oil change, (l/y);
ρ_{fuel} – fuel density, (kg/m^3) (Calculated according to Table 3 [18]);
LHV_{fuel} - lower heating value of combustion of fuel, (kWh/kg) (Calculated according to Table 3 [18]).

4. **Annual fuel consumption for the relevant category of vehicles after changing the engine oil:**

$$Q_A = F_A * D_a\ (l/y), \tag{4}$$

where:
Q_A – annual fuel consumption for vehicle category after engine oil change, (l/y);
F_A- average fuel consumption of all vehicles of the respective category after changing the engine oil (i=1,2,3), l/100 km, where:

$$F_A = F_B - F_B * f_r \ (l/y),\tag{5}$$

f_r – coefficient of reduced fuel consumption after changing the engine oil, % (Table 4)
D_a – average annual mileage of the vehicles of the respective category, (km/y)

5. **Energy delivered by fuel before engine oil change**

$$FE_A = Q_A * \rho_{fuel} . LHV_{fuel} \ (MWh/y),\tag{6}$$

where:
FE_A- Delivered energy with the fuel for the respective category of vehicles after changing the engine oil, (MWh/y);
Q_A – annual fuel consumption for vehicle category after engine oil change, (l/y);
ρ_{fuel} – fuel density, (kg/m^3) (Calculated according to Table 3 [18]);
LHV_{fuel} - lower heating value of combustion of fuel, (kWh/kg) (Calculated according to Table 3 [18]).

6. **Final energy saved**

$$FES = FE_B - FE_A \ (MWh/y),\tag{7}$$

where:
FES – final energy saved, MWh/y;
FE_B – energy delivered with the fuel for the respective vehicle category before the engine oil change, (MWh/y);
FE_A – Delivered energy with the fuel for the respective category of vehicles after changing the engine oil, (MWh/y);

7. **Savings of primary energy**

$$PES = (FE_B - FE_A). e_p \ (MWh/y),\tag{8}$$

where:
PES – primary energy saved, MWh/y;
FE_B – energy delivered with the fuel for the respective vehicle category before the engine oil change, (MWh/y);
FE_A – Delivered energy with the fuel for the respective category of vehicles after changing the engine oil, (MWh/y);
e_p – coefficient accounting for the losses of extraction/production and transmission of energy and energy resources ((Found according to Table 2 [17])

Carbon dioxide emissions saved

1. **CO_2 savings**

$$CO_2 = FES. f_i (\frac{t_{CO_2}}{r}),\tag{9}$$

where:
FES final energy saved, MWh/y;
f_i – Emission factor (t_{CO_2}/kWh) (Calculated according to Table 1 [17]).

3 Results and discussion

Based on the studies performed in the field of reducing greenhouse gases and CO_2 emissions and the developed methodology for evaluating the energy savings when changing the engine oil of vehicles, a calculation example is presented in Table 5, Table 6 and Table 7. The maximum mileage for changing the oil is 10,000 km. In Table 4 are presented the four categories of road vehicles, which were considered in the presented study:

Table 4. Vehicle categories.

№	Category	Type of vehicle	Weight	Number of passengers
1	M1.1	Private	3.5 tons	Up to 6
2	M1.2	Company	3.5 tons	Up to 6
3	M2	Cars up to 3.5 tons	3.5 tons	6-8
4	N1	Light trucks up to 3.5 tons	3.5 tons	-

Table 5. Calculation example – part 1.

№	Type of road vehicle	Type of fuel	Vehicles with changed oil, only EURO 4.5.6	Average fuel consumptio n before oil change	Average mileage of a car in category	Average mileage of a car for oil change	Average annual mileage for all cars
	M type	Fuel (-)	N number	F_B l/100 km	L km/y	D km/y	D_a km/y
1	M1.1	gasoline	665 332	7.94	11 067	11 067	7 252 559 224
2	M1.2	gasoline	62 130	8.37	20 399	20 399	1 267 389 870
3	M2	gasoline	794	11.10	12 308	12 308	9 772 332
4	N1	gasoline	304	10.51	20 278	20 278	6 164 389

Table 6. Calculation example – part 2.

№	Annual fuel consumption per category before oil change	Energy delivered with the fuel	Average fuel consumption of a car of a category after an oil change	Annual fuel consumption after oil change	Energy delivered with the fuel after an oil change
	Q_B l/y	FE_B MWh/y	F_A l/100 km	Q_A l/y	FE_A MWh/y
1	575 853 204	5 237 791	7.74	561 456 874	5 106 846
2	106 080 532	964 877	8.16	103 428 519	940 755
3	1 085 172	9 870	10.83	1 058 043	9 624
4	647 991	5 894	10.25	631 791	5 747
Total	**683 666 899**	**6 218 432**		**666 575 226**	**6 062 972**

Table 7. Calculation example – part 3.

№	Net final energy	Final energy saved	Primary energy saved	Emissions saved
	FE_A MWh/y	FES MWh/y	PES MWh/y	CO_2 t/y
1	5 106 846	130 945	144 039	34 438
2	940 755	24 122	26 534	6 344

3	9 624	247	271	65
4	5 747	147	162	39
Total	**6 062 972**	**155 461**	**171 007**	**40 886**

After applying the methodology for evaluating the energy savings at changing the engine oil of vehicles in the calculation example, the following results were obtained: for the annual amount of final energy saved (FEStot) – 155 461 MWh/y; for the annual amount of saved primary energy (PEStot) – 171 007 MWh/y and for the amount of saved carbon emissions (CO2) – 40 866 t. The results of applying the methodology are applicable only to certain categories of cars and practice shows that they enable an easy process of monitoring and verification.

4 Conclusions

The developed methodology for evaluating the energy savings, when changing the engine oil of vehicles, makes it possible to calculate the energy consumption of the vehicle after the oil change. Changing the engine oil of the car covers the replacement of the old oil with a new one, within the maximum mileage determined by the manufacturer. The methodology is applicable only to vehicles of categories M1, M2 and N1, which are first registered after 2007, i.e. Euro 4, Euro 5 and Euro 6. In addition, the application defines the main steps and principles of the procedure for third-party proof of energy savings achieved. As a result of the considered calculation example, the following results were obtained: for the annual amount of saved final energy (FEStot) – 155 461 MWh/y; for the annual amount of saved primary energy (PEStot) – 171 007 MWh/y and for the amount of saved carbon dioxide emissions (CO2) – 40 866 t, which confirm the use of the methodology.

It can be concluded that the methodology allows the possibility of estimating the energy savings after the replacement of the vehicle's engine oil, to be calculated by using the methodology. This would help reduce greenhouse gas emissions and CO_2 emissions worldwide.

The authors would like to thank the Research and Development Sector at the Technical University of Sofia for the financial support.

References

1. K. Midor, T. N. Ivanova, M. Molenda, W. Biały, O. V. Zakharov, J. E **15**, 259 (2022)
2. J. Zhang, H. Spikes, *Tribol. Lett.*, **63** (2016)
3. H. Spikes, Friction Modifier Additives, *Tribol. Lett.*, **60** (2015)
4. K. Froelund, M. G. Ross, *SAE Technical Paper* 2005-01-3715 (2005)
5. M. T. Devlin, J. Lubric., **6,** 68 (2018)
6. Z. Liu, A. Gangopadhyay, S. Simko, W. Lam, M. T. Devlin, *SAE Technical Paper* 2018-01-0933 (2018)
7. V.W Wong, S. C. Tung, *J. Friction*, **4** (2016)
8. B. Tormos, L. Ramirez, G. Miro, T. Perez, *SAE Technical Paper* 2017-01-2353 (2017)
9. Ricardo/ED12057/Final Report/Issue Number **3**, *(*2019)
10. European Parliament, Directive 2012/27/US, *Journal of the European Union* (2012)
11. European Parliament, Directive 2018/2002, *Journal of the European Union* (2018)
12. N. Labanca, P. Bertoldi, *Publ. Off. of the Euro Union*, JRC9969 (2015)

13. M. Mihaylov, I. Stoyanov, T. Iliev, E3S Web Conf., **327** 04003, (2021), https://doi.org/10.1051/e3sconf/202132704003
14. Ministry of Energy, *Law for energy efficiency* (Sofia, Bulgaria, 2015)
15. ACEA, *Light Duty Oil Sequence General Requirements* (Brussels, Belgium, 2021)
16. ACEA, *European oil sequences* (Brussels, Belgium, 2016)
17. Ministry of Energy, *Regulation No.E-RD-04-3 of 05/04/2016* (Sofia, Bulgaria, 2021)
18. B. Elvers, A. Schutze, *Handbook of Fuels: Energy Sources for Transportation* (2nd ed. Wiley, 2021)

Experimental investigation of water turbine with oscillating blades

Rositsa Velichkova[*], *Martin* Pushkarov, *Iskra* Simova, *Detelin* Markov, *Ivan* Denev, and *Radostina* Angelova

Technical University of Sofia, 8 Kl. Ohridski Bld, Sofia, Bulgaria

Abstract. Energy from renewable sources reduces greenhouse gas emissions and lowers our dependence on imported fossil fuels. In the presented work, a new way of harnessing the energy of sea waves is given, namely a turbine with oscillating blades. The experimental stand on which the experimental tests were carried out at different revolutions of the hydromotor of the turbine and at four different angles of the blade is presented.

1 Introduction

Electricity is the main driving factor for the development of all branches of industry, transport, utilities and agriculture. It plays an ever-increasing role in the acceleration of technical progress, ensures high rates of labor productivity, improves working conditions, living conditions, exerts an enormous influence on the distribution of productive forces and, more than any other factor, determines the level of the economic development of the national economy and the national income of society.[1-2]

Forecasting the development of global electricity production for several decades is complicated by many factors. It is assumed that by the end of the century, the production of electrical energy will reach 20-25 trillion kWh worldwide, and according to other data, 30-35 trillion kWh [2]. The issues of effectively meeting energy needs and overcoming the global energy crisis are particularly relevant today. Proceeding from the real objectivity - reduction of energy-raw resources (oil, coal, natural gas and others), the tendency to solve the energy problem is imposed, at the expense of the rational use of one's own resources - taking into account the geographical, natural-economic specifics and the scientific and technical potential of each country [1].

Waves have the highest energy density of renewable energy sources, compared to others like wind, solar, biomass and geothermal. This means waves have the greatest potential to be an important contributor to the world's "energy mix resilience" [3,4].

In the works, an experimental study of a new hybrid system for the utilization of wave energy is made. This experiment is based on already done analytical studies in [5-6].

* Corresponding author: rvelichkova@tu-sofia.bg

2 Experimental test-rig

The turbines are driven by a hydraulic station, which drives a hydraulic motor, which in turn transmits the movement to the turbines. In the first case, by means of a steel rope coupled to the hydraulic motor and the eccentric roller, it simulates the movement of waves in the seas and oceans, thus the turbine with the oscillating blades rotates and generates its power. When the turbine with oscillating blades Figure 1 lowers and raises its blades, they change their angle of attack and thus utilize the energy of the wave. [7].

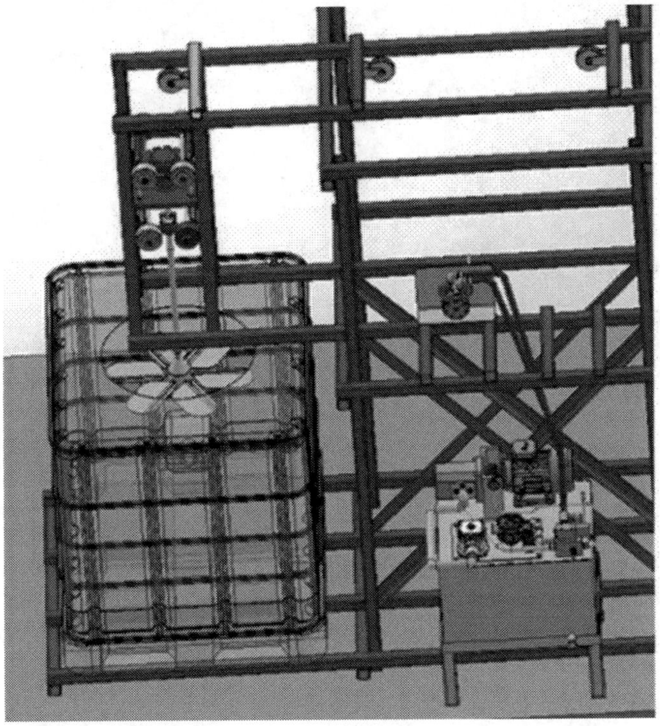

Fig. 1. Experimental test-rig.

2.1 Description of the experiment performed

The oscillating blade turbine experiment is to limit the working blades of the turbine to different angles.

The blades, being of the oscillating type, follow the amplitude of the wave and change their angle of attack. The operating range at the different wave levels that follow vary from (- 90° to 90°). The limitations of the angle of the blades of the turbine is presented in Figure 2.

In order to confirm the angle of attack of the blade through which the turbine has the highest revolutions respectively and torque of the developed bench, the following elements are adjusted.

Fig.2 Turbine with oscillating blades.

The oscillating vane turbine performs the wave amplitude using a hydraulic system to which a "crankshaft" is attached, which serves to change the wave amplitude in its experimental test. The rotation frequency of the crank is set by the hydraulic motor connected to the hydraulic system. In order to carry out the experiment, it is necessary to attach the following measuring devices to the stand, which will measure the quantities and parameters requested by us. A shaft speed or so-called wave frequency or amplitude sensor is mounted on the drive shaft of the hydromotor. A digital tachometer is attached to the shaft of the turbine with oscillating blades for measuring the revolutions of the turbine, a torque meter measures the torque realized by the turbine, as well as an additional brake to load the shaft of the turbine to measure the torque. In Figure 3 shows the system for measuring the parameters of the turbine shaft, where: 1 – Hydraulic motor;2 – Crank (Through it we determine the amplitude of the wave (It has the different sizes 50, 70, 90, 110, 150 mm));3 – Construction (or supporting frame of the elements);4 – Scale for load settings on the turbine; 5 – Flywheel;6 – RPM sensor; 7 – Coupler (between the turbine load and the torque sensor);8 – Torque meter ;9 – Coupler (between the torque sensor and the turbine shaft);10 – Cardan coupling (used so that the turbine shaft can be moved smoothly along the three axes);11 – Turbine shaft; 12 – Supporting frame of the test-rig;

Fig.3 Scheme of the measurement.

Each test shall be carried out according to the following procedure:

1. The hydraulic system is turned on;

2. The revolutions of the hydraulic motor are set to a certain value that we need (in this case, we have selected the following revolutions of the output shaft of the hydraulic motor of 10, 19 and 23.5 min-1);

3. After the adjustment, the measurement of the revolutions of the turbine shaft begins;

4. The torque measurement is visualized on the torque meter monitor;

5. The load (brake) of the turbine can be changed, and in the case of the test it is set to a certain load desired by us;

6. If necessary, the amplitude of the wave can be changed using the Crank.

3 Results

The initial data of the experiment are given at Table 1:

Table 1. Initial data of the experiment

$r_{ст}$	D_{1tur}	D_{2tur}	z	F_1	$r_в$	ρ	δ	$R_{кр}$
m	m	m	num	m^2	m	kg/m^3	mm	m
0.18	0.5	0.08	6	0.026	0.05	1000	12	0.09

The blade angle is 55^0,60^0,65^0,70^0. We choose this value due to our results at [7]. The average velocity of the turbine is determined by:

$$V_a = \frac{2R.n}{15} \tag{1}$$

The power of turbine is determined by:

$$N_T = 0,00145 z\rho F_1 \sin\alpha \left(V_a \sin 2\alpha\right)^3, kW \tag{2}$$

The torque of the turbine is determined by:

$$M_T = 0,4437 az\rho F_1 \sin\alpha \left(V_a \sin 2\alpha\right)^2, \tag{3}$$

where R - radius of the crank, m, n - revolutions,min $^{-1}$ z- number of blades, F_1 - face of blades, m^2, α -angle of the blades.

At Table 2, 4, 3 and 5 are given the results which is obtain from the experiments.

Table 2. Data at angle of the 55.

-	n_{hM}, min^{-1}	n_t, min^{-1}	Mt,N.m
1	10	7	1,01
2	19	35.1	1,08
3	23.5	44	1.146

Table 3. Data at angle of the 60

-	n_{hM}, min^{-1}	n_t, min^{-1}	Mt,N.m
1	10	8	1.1
2	19	35,6	1,18
3	23.5	48.5	1.285

Table 4. Data at angle of the 65

-	n_{hM}, min^{-1}	n_t, min^{-1}	Mt,N.m
1	10	9	1.23
2	19	39.1	1.4
3	23.5	51.2	1.41

Table 5. Data at angle of the 70

-	n_{hM},min^{-1}	n_t, min^{-1}	MtN.m
1	10	11	1.26
2	19	40	1.42
3	23.5	52	1.52

where n_{hm}-revolutions of hydro motor and n_t-revolutions of turbine.

Based on equation 1-3 is made calculation using the initial date given in Table 1 to 5.

t Figure 4 is given the distribution of torque at different angel of the blades.

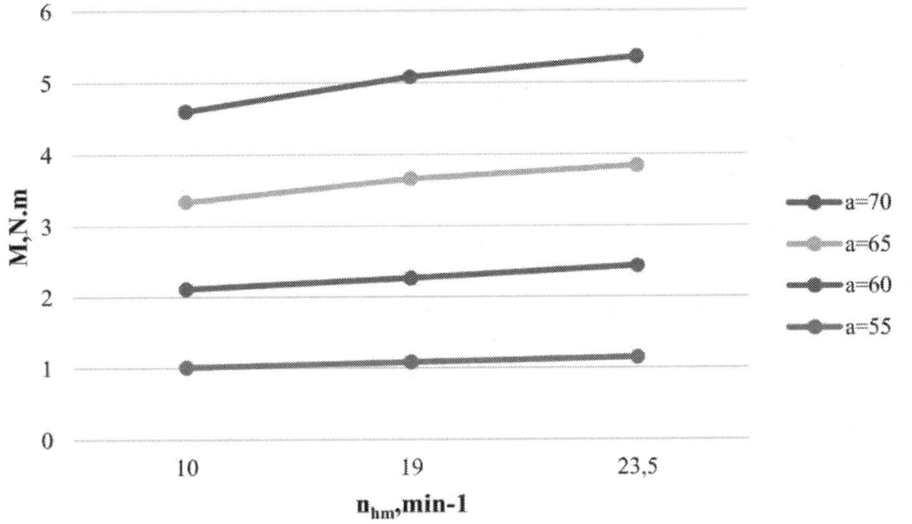

Fig. 4. Distribution of the torque.

At Figure 5 is given the distribution of the power of the turbine at different angle of the blades.

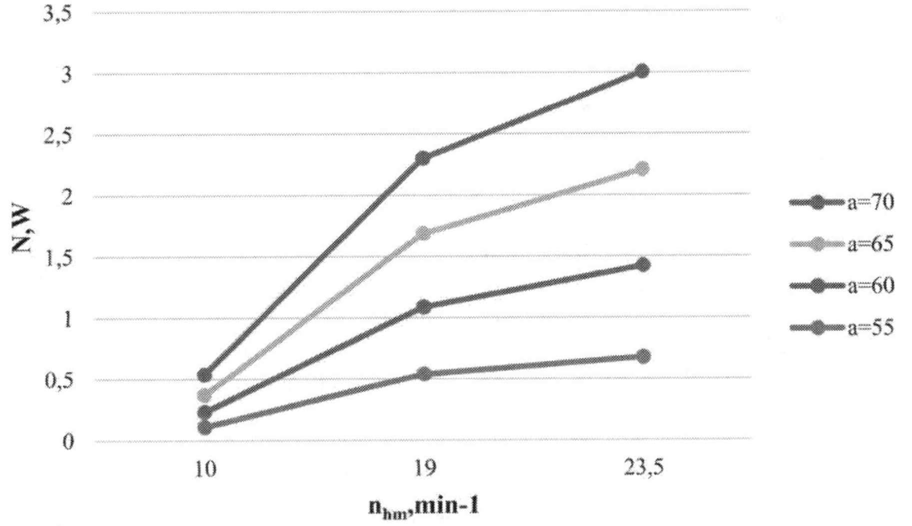

Fig. 5. Distribution of the power of turbine.

At Figure 6 is shown the efficiency at different angel of the blades.

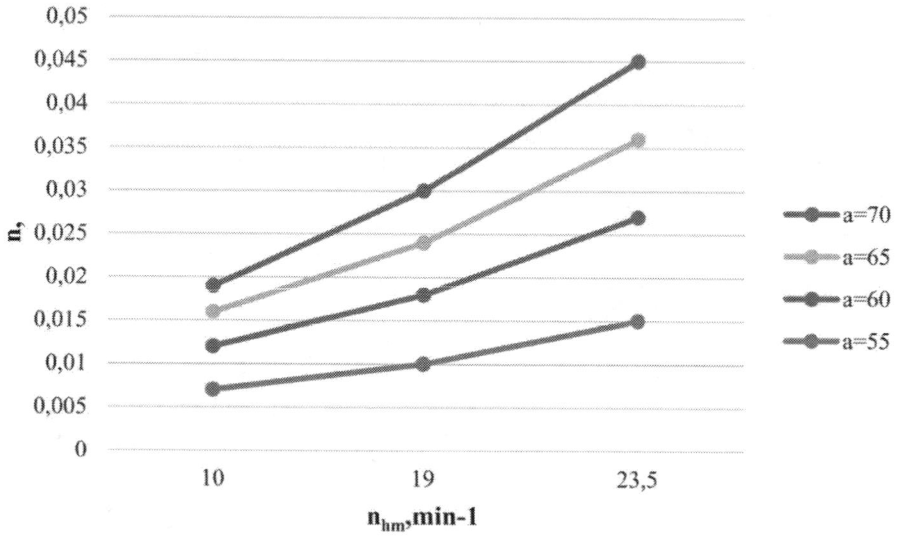

Fig. 6. Distribution of the efficiency.

From the obtained results, it can be seen that as the blade angle increases, both the torque and power of the turbine increase. In contrast, as the vane angle increases, the efficiency decreases. It can be seen that at angles 65^0 and 70^0 we have a much higher efficiency.

4 Conclusion

The presented paper presents experimental results of an oscillating blade turbine that harnesses the energy of sea and ocean waves. Several vane angles were investigated and these results are graphically presented. From the obtained results, it can be seen that the use of this type of turbine is quite effective and can help reduce the carbon footprint in nature. The results obtained from the analytical study [7] completely coincide with the data obtained from the experiment, which leads to the conclusion that the methodology used is effective and can be applied in practice.

This study is part of the project „Utilization of wave energy by hybrid system", KP06-H37/30, funded by Bulgarian Science Fund of the Ministry of Education and Science.

References

1. Ms. Neha, R. Joon, J. of. P **179,** (2021)
2. K.Kostov, I. Ivanov, K. Atanasov, EPh. And E. **2** (2021)
3. Y.Liu, B. Ding, B. Zhou, P. Cong,S. Zheng, F. in ER, (2020)
4. Ts. Tsalov, T., 7, (2017)
5. D. Markov, N. Valchev, R. Velichkova, I.Simova, Proc. 56th Sc. Conf. of RU, **6**, book 1.2, (2017)
6. R. Velichkova, I. Simova, D. Markov, V. Makakov M. Uzunova, M. Darchariff, G. Chen, *Analytical study of water turbine with fluctuating blades*, in Proceedings of the International Conference in Energy and Sustainability in Small Developing Economies (ES2DE), Funchal, Portugal, (2017)

A method for derating power of a MW direct-drive wind turbine permanent magnet generator as over allowable working temperature

Hung Vu Xuan*

Hanoi University of Science and Technology, School of Electrical and Electronic Engineering, Hanoi, Vietnam

Abstract. The use of direct-drive permanent magnet (PM) generators for MW wind application is increasing, because they have higher reliability and less maintenance than geared generators such as geared PM generators and geared induction generators. The price of the direct-drive PM generator is high, taking 10%-15% price of its overall system. Therefore, the generator mush be safely operated to have a long life. For a PM generator, its performance is sensitive to ambient and working temperature so that it needs to be monitored. In this paper, a method is proposed, combining the practical knowledge of a generator and theory to give permissible maximum generated power under certain ambient condition. The stator iron loss of a MW permanent magnet generator in direct-drive wind turbine application is presented as a function of rotor speed or frequency. The generator's copper loss and electromagnetic torque versus winding temperature and current are investigated. The relationship of voltage, current and active power to the generator temperature is addressed. The given performances can be used to choose the generator's working parameters such as voltage, current, output power for avoiding generator damaged by over temperature.

1 Introduction

The use of variable-speed direct-drive (gearless) permanent magnet (PM) generators for wind application is increasing, because they have higher reliability and less maintenance than geared generators such as geared PM generators and geared induction generators. Regarding geared generators, the gearbox components suffer tremendous stress due to wind turbulence and small defects in any component can bring the whole system to a halt. This makes the gearbox the highest maintenance part of a turbine. For off-shore wind turbines (WT), the gearbox parts face higher wind speeds and suffer more stresses compared to on-shore WT, and offshore turbines are so costly in terms of maintenance. Therefore, by eliminating the gearbox, the reliability of wind turbine system increases. A direct-drive generator is bigger and heavier than a geared generator because a direct drive generator runs at a lower rotor speed and a higher torque than a geared generator. As a result of its

* Corresponding author: hung.vuxuan@hust.edu.vn

high electromagnetic torque, electromagnetic force and high flux density required for high power density, the current direct-drive wind PM generator consists of complicated structures of manufacturing and uses expensive materials for the parts of which need magnetic characteristic (e.g., using iron laminations or using soft magnetic composites). The price of the direct-drive PM generator is however high, taking 10%-15% price of its overall system. Therefore, generator mush be optimally controlled and operated to ensure that generator has a long life [1-8].

How and when derate power of a wind turbine? There are two ways for derating power of a wind turbine. One is changing the rated rotor speed; this way is normally applied for a wind turbine installed at positions having either a lower wind speed or higher altitude and/or lower air density than the designed condition. The other way is changing the rated torque or power or electrical current; this way is normally implemented when the wind generator is installed in a place having either ambient temperature or altitude is over the specific condition in the design criteria, while the capacity of cooling system reaches its limit [9-11].

Why do utilize an availably designed WT generator rather than design a new wind turbine generator suitable to the condition at the wind turbine placing-position? This is because the new design and manufacturing of a wind turbine generator is really expensive. A new design of a component such as a cooler or retrofitting new equipment to improve cooling system is costly as well. Therefore, an available wind turbine designed in a specific condition is favorably adopted for new installations, except for in case of big amount of installations [10-11].

Over the designed temperature of a permanent magnet (PM) generator can affect on its life-span or damage the machine [7-8]. In calculation, the machine temperature can obtain from a thermal model [12]. In real operation, the machine temperature can obtain by getting the average value of some instant measurements. This paper proposes an efficiently de-rating strategy combining between practice and theory, and presents the performance of a 5 MW wind generator, which is one of biggest wind turbines currently installed in the author's own country. The performance can be used to properly choose current, voltage and output power of a generator according to working temperature. In other words, the performance can be used to adapt the power curve during operation at/above the rated wind speed according to working temperature. This helps for the generator working safety and having long lifespan. In practice, rated power of a wind turbine currently reaches over 15 MW, however the proposed method is still valuable for such a range of supper high power wind turbines.

The direct-drive PM generator is used for the analysis in this paper for the offshore wind application, having rated generator speed of 18 rpm. The generator is the type of inner rotor and outer stator. The stator is cooled by thermal convection with wind blowing through heat sinks mounted on the generator's stator yoke. Moreover, the rotor is cooled by air forced flow, in which the inlet air flow coming in one end side of generator, next air flow through the generator air gap to the other end side of generator, then through outlet pipe and finally releasing in the wind turbine's nacelle.

This paper is organized as follows. Firstly, the generator's iron loss is presented as a function of rotor speed. Next, the generator's copper loss and electromagnetic torque and current versus winding temperature are investigated. Last, the relation of voltage, current and permitted active power to the generator temperature is performed.

2 Stator iron loss versus rotor speed

In Figure 1, the generator's stator iron loss is presented as a function of rotor speed. The stator iron loss increases when rotor speed increases. As can be seen, the stator iron loss is about 28 kW at a rated rotor speed of 18 rpm. The formula to calculate stator iron is described as follows [13-14]:

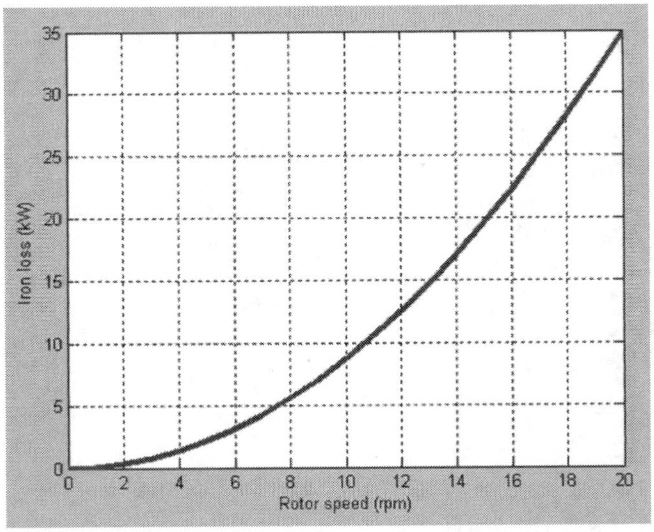

Fig. 1. Stator iron loss versus rotor speed of the studied generator.

$$p_{iron} = p_h + p_e = c_e \left[P_{Feoh} \left(\frac{B}{B_o} \right)^{\beta} \left(\frac{\omega}{\omega_o} \right) + P_{Feoe} \left(\frac{B}{B_o} \right)^2 \left(\frac{\omega}{\omega_o} \right)^2 \right], \quad (1)$$

where, p_h and p_e are the hysteresis and eddy current loss respectively. P_{Feoh} and p_{Feoe} are hysteresis and eddy current loss per kg of stator core material under tested condition with frequency f_o and flux density B_o, [W/kg]. C_e is a factor, accounting for that flux density in air gap is not sinusoidal, as the result in an extra increase of the iron losses in the PM generator with higher frequencies. β is the Steinmetz constant ($1.5 < \beta < 2.3$), which depends on the iron lamination material and thickness. Choosing β is equal to 2 in this study.

3 Resistive loss of stator winding versus winding temperature and current

Copper loss due to current flowing through wire is proportional to the resistance of the wire and the square of the current. Copper loss is calculated as:

$$P_{Cu} = mR_s I_1^2 \, . \quad (2)$$

In (2), m is number of phases, I_1 is the rms current running in the winding of a phase, R_s is the winding resistance of a phase. Note that resistance is a function of temperature as:

$$R_{s(T)} = R_{s(20)}[1 + \alpha_{Cu}(T - 20)] \, , \quad (3)$$

where, $R_{s(T)}$ is stator resistance at temperature T, $R_{s(20)}$ is stator resistance at temperature of 20°C, and α_{Cu} is the temperature coefficient of copper resistance.

Figure 2 shows the relation of copper loss to the winding temperature and the rms current of generator. It can be seen that when the winding temperature increases, the resistance of winding and copper loss increase. At a rated current and winding temperature of 90 °C, the calculated copper loss of the generator is 184.4 kW. This value is validated by measurement. It should be noted that the displacement current in windings due to the proximity effect and the skin-face effect increases the resistive loss so that a factor of 1.13 getting from finite element model is taken in account for the additional loss.

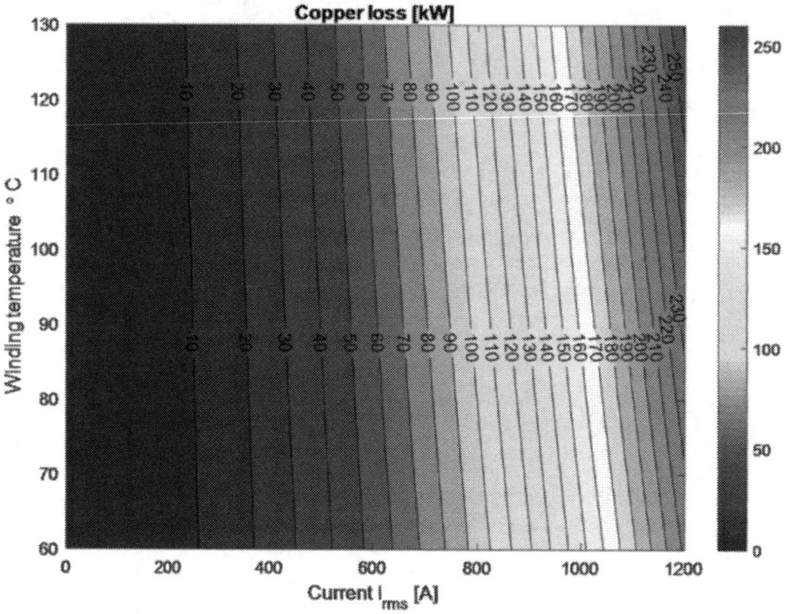

Fig. 2. Copper loss versus winding temperature and current of the studied 5MW-generator.

4 Torque of generator versus winding temperature and current

The formula to calculate electromagnetic torque is expressed as:

$$T_e = \pi r_s^2 l_s \hat{B}_{g1} \hat{K}_1 \sin\beta \,, \tag{4}$$

where β is the angular displacement between the field produced by the magnet and the stator current, and K_1 is the amplitude of the current sheet (electrical loading), B_{g1} is amplitude of flux density in air gap, r_s is radius of stator, l_s is stator length [15].

In the (4), the flux density in the air gap is calculated as the following equations [15]:

$$\hat{B}_{g1} = \frac{4}{\pi} B_{g,0} \sin(\alpha) \,, \tag{5}$$

$$B_{g,0} = \frac{l_m}{l_m + \mu_{rm} g_{eff}} B_r \,, \tag{6}$$

where, 2α is pole arc, l_m is the length of magnet in the radial direction, μ_{rm} is the relative permeability of the magnet, and g_{eff} is the effective air gap.

The torque-producing component of electric loading (assuming this component is the first harmonic component) can be written as:

$$K_1(\theta_s) = \hat{K}_1 \sin(p\theta_s - p\omega_m t + \beta) , \tag{7}$$

The amplitude of the current sheet (electric loading) is expressed as:

$$\hat{K}_1 = K_{w1} \frac{m(2N_w)\hat{I}_1}{2\pi r_s} , \tag{8}$$

where p is number of pole pairs, I_1 is the stator current, N_w is number of winding turns per phase. For further information, can find it in [15].

Figure 3 shows electromagnetic toque versus magnet temperature and the rms current. It is assumed that the power factor is constant and equal to 0.942 (measured). Rotor speed is 18 rpm. It can be seen that when the magnet temperature increases (the remanent flux density decreases), electromagnetic torque reduces. It is clear that the higher generator current, the higher electromagnetic torque.

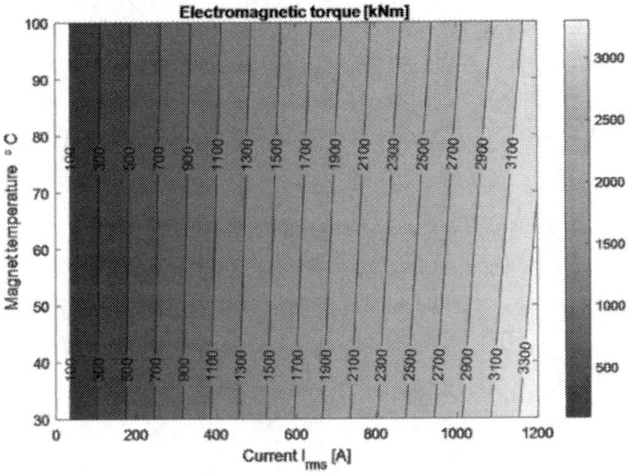

Fig. 3. Electromagnetic torque versus magnet temperature and current.

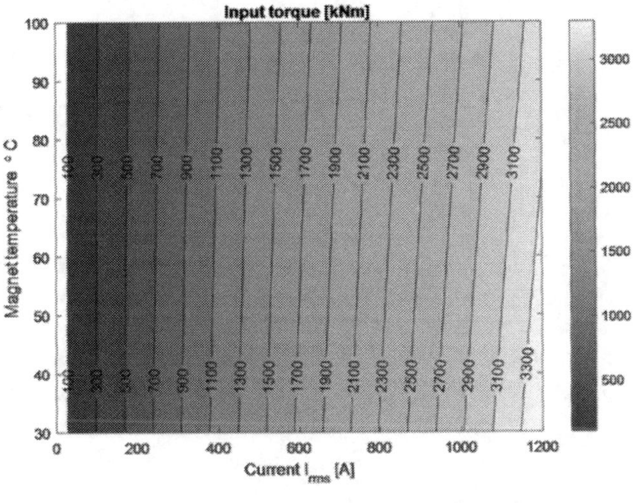

Fig. 4. The input torque T versus magnet temperature and current.

The input torque T presented in Figure 4 is defined as follows. The input torque is equal to electromagnetic torque T_e plus the torque T_0 corresponding to iron loss of the generator;

i.e., $T=T_e+T_0$. Clearly, $T_e>>T_0$ so that $T_e \approx T$. It should be noted that the input mechanical torque is equal to the input torque plus the torque due to the friction of the seal and bearing.

5 Derating generator power according to winding temperature

When the winding temperature increases, the winding resistance increases so that copper loss also increases. In addition, when magnet temperature increases, the remanent flux density of magnets decreases. This leads to the decrease of generator voltage. Therefore, if the output power of the generator is kept constant, the current of generator increases when the voltage decreases. To avoid overheat of the generator, the output power of generator should be reduced reasonably according to the winding temperature. It means that the generator current needs to be reduced to avoid over temperature of stator windings. In another case of the winding temperature lower than the limit temperature, the generated power can increase higher than the power on the designed power curve. In this section, the voltage, current and power of the generator are presented as a function of the temperature. Based on results, the working temperature and the limited output power of the under studied generator can be predicted.

The influence of temperature on the magnet and winding is presented by the following governor equations. Magnet flux density B_m relates to other parameters at room temperature as follows:

$$B_m = \mu_0 H_m + J \quad or \quad B_m = \mu_0 \mu_r H_m + B_r . \tag{9}$$

The remanent flux density B_r decreases with temperature. This effect is usually specified in terms of the reversible temperature coefficient of B_r. The remanent flux density at temperature T°C can be written:

$$B_{r(T)} = B_{r(20)}[1+\alpha_{Br}(T-20)/100] . \tag{10}$$

where $B_r(20)$ is the value of B_r at 20°C. α_{Br} is the reversible temperature coefficient of Br in percentage per Celsius degree; α_{Br} in range of -0.03 to -0.18 (%/°C).

To explain the calculation, let start with a simplified thermal model of generator as shown in Figure 5.

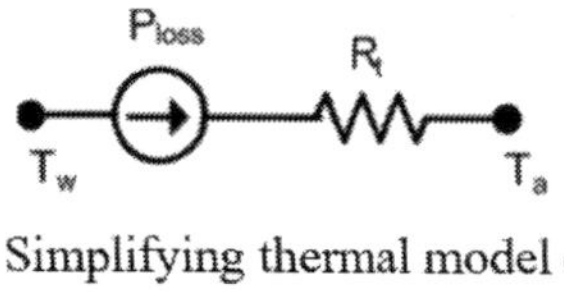

Fig. 5. Simplifying thermal model of generator stator.

In Figure 5, T_a is the ambient temperature, T_w is the winding temperature, R_t is thermal resistance of the generator and P_{loss} is the stator power loss.

If we assume that the heat produced by the rotor iron loss and mechanical iron loss is released to the ambient air by the air-forced-cooling system. The stator iron loss and copper loss are released to the ambient air through the heat sink mounted on the outer surface of stator. Therefore, the thermal resistance R_t can be calculated as:

$$R_t = (T_w - T_a) / P_{loss} . \tag{11}$$

If R_t and P_{loss} are known, the winding temperature can be calculated as:

$$T_w = T_a + R_t . P_{loss} . \tag{12}$$

Alternatively, the winding temperature can get from online measurement. If replace winding temperature T_w in (12) by the permitted winding temperature, the limited generator power loss can be calculated, then the limit of current can be calculated as well.

For example, from the calculation and measurement of the generator at a rated wind speed, a rated rotor speed of 18 rpm and rated current we have T_a=10 °C, T_w=91°C, $P_{loss} = P_{copper}+P_{iron}$=212 kW at 91°C, thermal resistance of 0.38 °C /kW. Therefore, if T_a=40 °C and the power loss P_{loss} is still the same; i.e., 212 kW, then temperature of winding is roughly predicted around 121 °C. In addition, when the generator temperature increases, the copper loss also increases. To avoid over the winding temperature (assumedly limited at 115 °C) and magnet demagnetization, the output power and current of generator should be reduced according to the increase of the generator temperature. The calculation shown in Figure 6 and Figure 7 is done with input parameters as:

- The power factor is constant and equal to 0.942 (from measurement of a 5MW-PM generator).
- Rotor speed is 18 rpm.

Figure 6 shows remanent flux density of magnet and winding resistance with respect to winding temperature. Note that copper loss dominates the total stator loss of generator. The calculated results of voltage, current and output power of the generator are presented in Figure 7. When the temperature increases, the winding resistance increases (Figure 6b) and the generator current should be decreased (Figure 7b) to reduce copper loss.

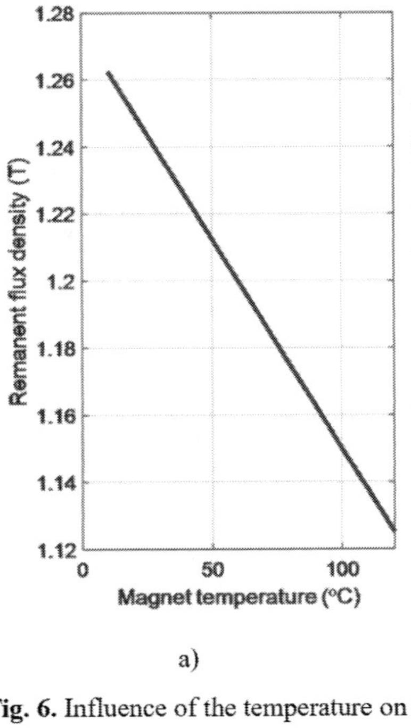

a)

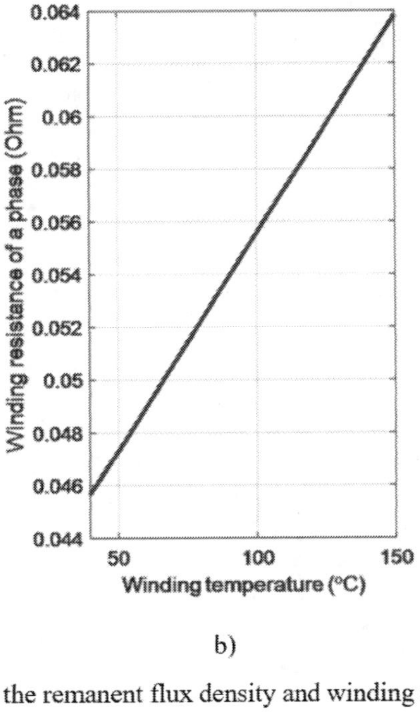

b)

Fig. 6. Influence of the temperature on the remanent flux density and winding resistance.

It can be seen from Figure 7 that at the working temperature of stator winding of 91 oC, line to line rms voltage is about 3048 V, the rms current is about 1067.9 A, the active power of generator is 5.27 MW. However, with the ambient temperature of 40 oC and the limited working temperature of stator winding at 115 oC, line to line rms voltage is 2950V, the rms current should be limited at 1035 A, the active power of generator should be limited at 4.95 MW.

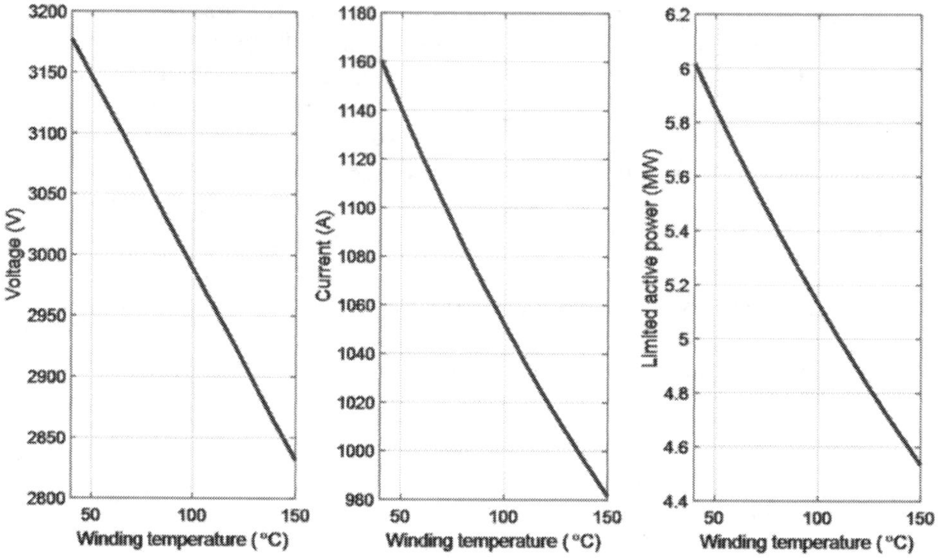

a) b) c)

Fig. 7. Line to line voltage, the limit of rms current and permitted active power of the generator with respect to temperature of stator winding.

Therefore, an average derating factor or derating rate of power for high ambient temperature is 2% for each 10oC increasing or 10.7 kW/°C. Based on Figure 7b, c, the limit of current and output active power with respect to winding temperature can be found. This is necessary to operate a wind turbine safely.

6 Conclusion

To avoid overheat of a generator when environment temperature increases over 40 degrees Celsius in summer season or when a generator is placed at a high-altitude or low pressure place and to maximize generating power at low ambient temperature, the output power of generator must be adapted according to the working temperature of generator. This paper has been presented a method to perform the relationship of generator's current, voltage and power to working temperature. The proposed method has combined the practical knowledge of a generator and theory to give allowable maximum generated power under certain ambient condition. This helps operate a MW wind turbine safely and efficiently. It is shown from this research that the derating factor or derating rate of power for high working temperature is 0.2%/oC or 10.7 kW/oC.

This research is funded by the Hanoi University of Science and Technology (HUST) under project number T2020-SAHEP-002.

References

1. H. Polinder, F. F. A. van der Pijl, G. -J. de Vilder, P. J. Tavner, IEEE Transactions on Energy Conversion, **21** (3), pp. 725-733, (2006)
2. H. Polinder, J. A. Ferreira, B. B. Jensen, A. B. Abrahamsen, K. Atallah, R. A. McMahon, IEEE Journal of Emerging and Selected Topics in Power Electronics, **1** (3), pp. 174-185, (2013)

3. D. J. Bang, H. Polinder, G. Shrestha, J. A. Ferreira, *Promising Direct-Drive Generator System for Large Wind Turbines*, in Proceedings of 2008 Wind Power to the Grid - EPE Wind Energy Chapter 1st Seminar, Delft, Netherlands, 2008, pp. 1-10, (2011)
4. S. Wang, A. R. Nejad, T. Moan, Wind Energy, **23** (4), pp. 1099-1117 (2020)
5. C. E. S. de Souza, E. E. Bachynski-Polić, Marine Structures, 84, 103182, (2022)
6. S. Rajendran, M. Diaz, R. Cárdenas, E. Espina, E. Contreras, J. Rodriguez, Processes, **10** (11), 2302, (2022)
7. IEEE Recommended Practice for Emergency and Standby Power Systems for Industrial and Commercial Applications, in IEEE Std 446-1995 (The Orange Book), vol., no., pp.1-320, 3 July 1996, doi: 10.1109/IEEESTD.1996.85950.
8. IEEE Standard Definitions for Use in Reporting Electric Generating Unit Reliability, Availability, and Productivity, in IEEE Std 762-2006 (Revision of IEEE Std 762-1987), vol., no., pp.1-75, 15 March 2007, doi: 10.1109/IEEESTD.2007.335902.
9. D. Astrain Juangarcia, I. Eguinoa, T. Knudsen, J. Phys.: Conf. Ser. **1037** 032039, (2018)
10. Q. Xiong, et al, Controlling the temperature of a wind turbine electric generator by varying the electrical power factor, European patent, EP 2 060 786 A2, 20 May (2009)
11. B. H. Kristian, et al, Derating a wind turbine as a function of wind speed, WIPO, WO/2021/004591, 14 Jan (2021)
12. H. V. Xuan, *Guidelines of Effective Thermal Modeling of an Outer Rotor PM Machine with Concentrated Windings*, in Proceedings of the 2022 International Conference on Communications, Information, Electronic and Energy Systems (CIEES), Veliko Tarnovo, Bulgaria, 2022, pp. 1-7, doi: 10.1109/CIEES55704.2022.9990783.
13. H. V. Xuan, D. Lahaye, S. O. Ani, H. Polinder, J. A. Ferreira, *Electrical generators for maritime application*, in Proceedings of the 2011 International Conference on Electrical Machines and Systems, 2011, pp. 1-6, doi: 10.1109/ICEMS.2011.6073526.
14. H. V. Xuan, S. O. Ani, D. Lahaye, H. Polinder, J. A. Ferreira, *Validation of non-linear dynamic FEM model for design of PM machines with concentrated windings in ship application*, in Proceedings of the 2011 14th European Conference on Power Electronics and Applications, 2011, pp. 1-9.
15. Hung Vu Xuan, Chapter 3, Modelling of outer rotor PM machines with concentrated windings, Ph.D, thesis, (TU Delft, The Netherlands, 2012)

Carbon footprint of reservoirs in Bucharest

Gabriela Elena Dumitran, *Liana Ioana* Vuță, *Angela* Neagoe, *Eliza-Isabela* Tică and *Bogdan* Popa*

University Politehnica of Bucharest, Hydraulics, Hydraulic Machines and Environmental Engineering Department, Faculty of Energy Engineering, 060042, Bucharest, Romania

Abstract. The paper presents the carbon footprint of the 10 reservoirs on Colentina river in Bucharest. There were presented entry data and hypotheses used by G-res tool who was applied for determining GHG emissions of these reservoirs. Therefore, we now have a good picture about their contribution to the overall GHG emissions in Bucharest.

1 Introduction

Used since ancient times, the regulation of water courses and the creation of lakes, with numerous benefits for the population - irrigation, water supply, navigation, flood protection, hydroelectricity, recreation - are common both in and near cities and in remote, hard-to-reach areas. Today few rivers are untouched by human activities.

The creation of freshwater reservoirs by damming rivers results in changing the characteristics of the aquatic system from a flowing to a static, lentic system. These lentic systems are characterized by active carbon exchange from one species to another, mineralization at different end products, enhanced sedimentation, gas emissions at the air-water interface.

Concerns about climate change generated research to determine the contribution of natural or anthropogenic lakes to greenhouse gas emissions, the variation of this contribution in different regions of the world and under different environmental conditions, to determine the ecological footprint of newly created or existing reservoirs [1-3].

As a result, numerous studies in recent years confirm that inland waters, regardless of their type - stream/river, pond/lake/reservoir - represent an important component of the total greenhouse gas (GHG) emissions budget [4-12].

Estimating greenhouse gas emissions from lakes is a complex problem, and recent research has developed tools and methodologies, as G-res, to assess it [1-3, 13-17]. Basically, all the methodologies are based on measurements of GHG emissions from reservoirs in different climate zones. The reported results highlight the importance of certain characteristics of reservoirs/watercourses and their catchments, e.g. climate, reservoir age and depth, watershed lithology and land cover, water quality – pH, dissolved oxygen, TN, TP, trophic state, and so on [18-21].

In this paper, GHG emissions related to the cascade of reservoirs on the Colentina River, in Bucharest area, are evaluated by using G-res tool.

* Corresponding author: popab_234@yahoo.com

2 Materials and methods

2.1 G-RES tool

Available online, for free, G-res allows users to determine GHG emissions of natural or artificial lakes and reservoirs. It also presents the advantage of performing the allocation of GHG emissions from reservoirs between the different uses, as well as the estimation of the possible emissions associated with different unrelated anthropogenic sources.

It can be used for the study of existing lakes or for those expected to be created and considers the emissions due to the construction works performed. As input data, values related to the climate, the age of the reservoir, solar radiation, the water uses, the land cover, the type of soil, the surface of the reservoir and the hydrographic basin, as well as the concentration of phosphorus in the reservoirs and the carbon content of the soil are required.

As a result, the emissions before and after the development are delivered, including the evolution over 100 years, the estimated lifetime of the reservoir, the emissions from non-anthropogenic sources (UAS), the emissions from the construction phase and the allocation of GHG emissions between the different uses of the reservoir [14].

2.2 Description of the study area

The Colentina River is a typical small, heavily meandering river of the countryside, which naturally often dries up in summer. It is part of the Argeş river basin and is a left tributary of the Dâmboviţa River, which in turn is a tributary of the Argeş River [22, 23]. The river basin of Colentina River is presented in Figure 1 and some hydrographical and hydrological characteristics are presented in Table 1 [24].

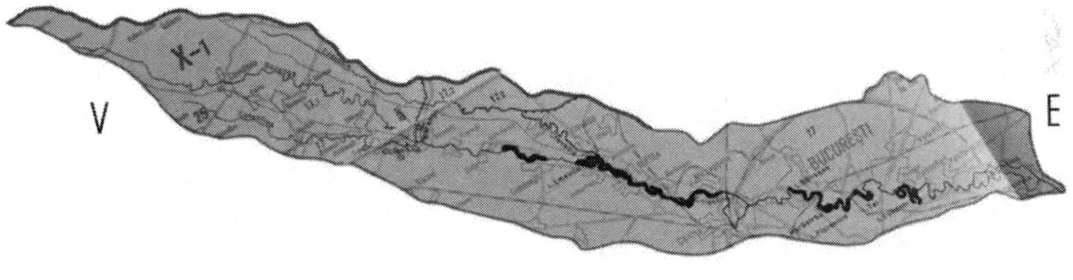

Fig. 1. Colentina River hydrographic basin [25].

Table 1. Characteristics of the Colentina River [24].

Source Area	Dâmboviţa county, Nucet village
Mouth	Dâmboviţa, Cernica village
Altitude at the Mouth	52 masl
Altitude Difference	127 m
Catchment Area	643 km^2
Length of Watercourse	101 km
Average Flow	0.60 m^3/s – Colacu (upstream Bucharest) 2.39 m^3/s – Cernica (downstream Bucharest)

Colentina is a river body of indigenous origin, which springs from the southern end of the alluvial cone at Târgovişte, in the village of Nucet, at an altitude of 179 masl. After crossing 101 km, of which 37.4 km are in the Bucharest area, it flows into the Dâmboviţa

River. In its course, Colentina River collects the flows of two tributaries: Baranga stream, with a length of 31 km, and Crevedia stream, 30 km long. These tributaries together with their collector drain cover an area of 643 km^2.

The density of the hydrographic network in the Colentina basin is 0.25 km/km^2, below the national average. The sinuosity coefficient of the riverbed is of 1.56 and its slope is less than 1‰. In the middle and lower reaches the degree of meandering is very pronounced, leading to extremely low water velocities.

Data recorded at the Cernica hydrometric station show that a mean flow of 2.39 m^3/s is transited through the Colentina River. During summer, and especially during dry periods, the lowest flows are recorded, up to 8-10 l/s. The average specific runoff calculated over the entire basin is 3.8 $l/s/km^2$. In rainy years, exceptional flood waves may occur, as happened in July 1975, when heavy rainfall recorded the historical maximum flow of 57.9 m^3/s (Colacu station). The low slopes and low flow velocity favored the creation of marshy areas, which had to be drained and transformed into reservoirs [26].

The general action plan for the hydraulic development of the Bucharest region was drawn up in 1932 by the engineer Nicolae Caranfil and included several stages:

- **Stage I**: Buftea reservoir; 18 months, completion September 1934,
- **Stage II**: Băneasa and Herestrău reservoirs; 5 months, completion June 1935,
- **Stage III**: Bilciurești dam on the Ialomița River and the Bilciurești - Ghimpați diversion; 18 months, completion November 1936. Floreasca reservoir, completion in September 1936,
- **Stage IV**: Tei reservoir (including the dam); 16 months, completion September 1937,
- **Stage V**: Fundeni reservoir; completion 1938-1939. Mogoșoaia-Băneasa channel; completion 1939-1940,
- **Final stage**: Pantelimon and Cernica reservoirs, with the capital's industrial port and the link with the Dâmbovița Argeș navigation channel, completion 1941-1950 [27].

By 1940 only 7 of the 15 reservoirs had been completed. The following ones were built from 1968 onwards, with some modifications to the original plans. In 1970 the whole series of reservoirs on the Colentina River was completed as they exist today.

Nowadays there are 15 reservoirs on the Colentina River, about 56 km long and a total head of about 49 m, with a total surface area of 1500 ha and a total live storage of 44 Mm^3, created by damming the river and draining the marshy/swampy riverbed. The dams are of the homogeneous earth dam type, with the upstream side covered with concrete slabs and the downstream side covered by grass. Of these reservoirs, 10 are in the territory of Bucharest, marked in Table 2 (Străulești to Dobroești), where water uses means: fish farming (F), irrigation (I), flood protection (FP), recreation (R).

Along the Colentina River there are no dykes. The flood defense infrastructure consists of 18 dams on the upstream side, which control the flow of the Colentina River, Figure 2.

On the tributaries there are about 20 dams more. The dams on the Colentina River are either for fish farming or for flood protection, starting from upstream: 2 dams (Ciocănești 1 and Ciocănești 2) with low attenuation volume are under the administration of the fishing companies; The other 16 dams are under the administration of NARW (National Administration Romanian Waters) (7) and the municipality of Bucharest (9). The largest, Buftea reservoir, located furthest upstream, has a large attenuation volume and controls most of the flood flow. All reservoirs belong to the Bucharest flood defense line [29].

During summer, reservoirs on Colentina River show the phenomenon of algal blooms due to the high concentration of phosphorus and nitrogen in the water, substances that enter the water through several pathways, including insufficiently treated domestic wastewater from upstream communities along the river or from the supplemental food offered to fish in reservoirs where fish farming is practiced [24].

Table 2. The 15 reservoirs on Colentina River [28].

Reservoirs	Surface [ha]	Max depth [m]	Volume [Mm^3]	Water uses*
Buftea	307	8	9.6	F, I, FP, R
Mogoşoaia	66	4	2	F, I, FD, R
Chitila	75	5	0.86	I, R, F, FP
Străuleşti	**39**	**5**	**0.7**	**F, I, FP, R**
Grivițta	**80**	**4**	**1**	**F, I, FP, R**
Băneasa	**40**	**3**	**0.6**	**F, I, FD, R**
Herăstrău	**77**	**5**	**1.5**	**F, I, FP, R**
Floreasca	**70**	**5**	**1.6**	**F, I, FP, R**
Tei	**80**	**4**	**1.9**	**F, FD, R**
Plumbuita	**55**	**4**	**0.9**	**F, I, FD, R**
Colentina	**29**	**5**	**0.6**	**F, I, FD, R**
Fundeni	**123**	**5**	**8**	**F, I, FD, R**
Dobroeşti	**120**	**2**	**2**	**F, I, FD, R**
Pantelimon	260	4	12.3	F, I, FD, R
Cernica	360	3	0.44	R, F, I, FD

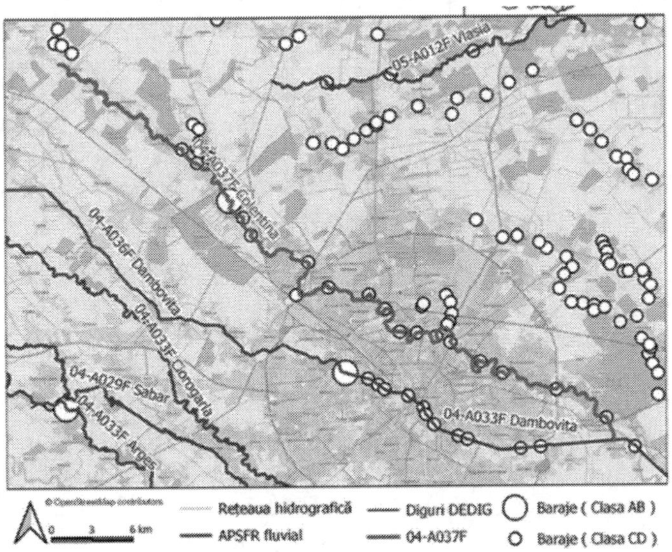

Fig. 2. Flood defense infrastructure of the Colentina River.

These reservoirs do not face serious problems of contamination with pollutants, but even after a brief analysis of some water parameters, the presence of pathogenic germs can be observed, which indicates that the water has been contaminated with human or animal faecal matter and shows the need to connect to the sewage network some marginal neighborhoods or small upstream localities. It also shows that the maximum permissible values for cadmium, copper and lead, heavy metals which can be a danger to human health if water is consumed directly from the river or if fish is eaten excessively, are exceeded, as they tend to accumulate in animal tissues and eventually end up in the human body.

At present, the water quality of the Colentina River's reservoirs is inadequate. This is due to the direct discharge of wastewater upstream from Bucharest into the Colentina River by industrial units and the population: Buftea (food industry, light industry, population, pollution consisting of nutrient inputs), Crevedia (poultry farm, population), Mogoşoaia (film industry, population). On the other hand, the reservoirs on the Colentina River have not been dredged for over 30 years, leading to the accumulation of large quantities of insalubrious sludge.

In this case study, only the 10 reservoirs included in the territory of Bucharest municipality were considered, and the entire Colentina river basin as the watershed. To apply G-res on only an equivalent reservoir, the 10 reservoirs were considered as forming a single water body, whose characteristics are obtained by summing volumes and surfaces and by averaging depths.

Regarding the population at the level of the hydrographic basin, 2 M inhabitants were considered for Bucharest and 1.5 M for the neighboring areas. As uses of this water body were considered: recreation - primary, flood mitigation - secondary and irrigation - tertiary, the degree of allocation of water volumes related to these uses being implicitly defined in the G-res.

As regards land cover, for Bucharest the values are those reported in the cadastral maps, while for the neighboring areas [30], where data were not available, they were estimated based on satellite maps. The centralization of these data is shown in Table 3.

Table 3. Data for land cover in the Colentina river basin.

Land cover in the catchment area	Bucharest 36.55%	Neighboring areas 63.45%	Total 100%
Croplands	11	50	30.5%
Grassland	2	5	3.5%
Forest	21	20	20.5%
Water bodies	4	4	4%
Bare areas	0	0	0
Settlements	61	20	40.5%
Wet areas	0.4	0.5	0.45
Drained Peatlands	0.6	0.5	0.55
Total	100%	100%	100%

2.3 Data used in G-res tool

To perform GHG emission analysis, G-res tool needs annual mean meteorological data (average monthly temperatures, average wind speed, and mean global horizontal radiance), which have been retrieved, for Bucharest area, from www.meteoblue.com, a meteorological free data service, Figure 3.

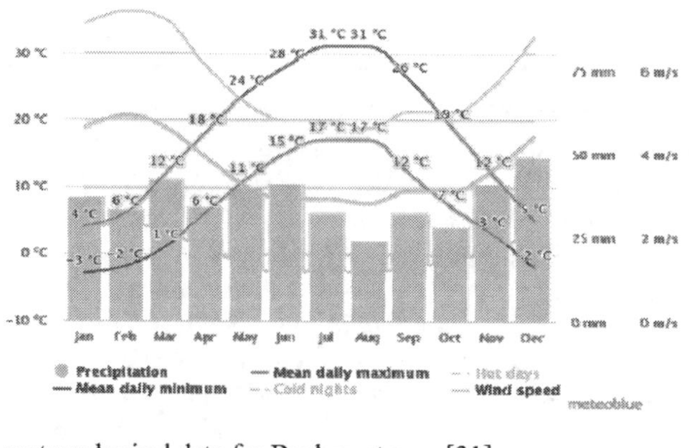

Fig. 3. Annual mean meteorological data for Bucharest area [31].

As regards the other data needed by G-res tool, the values used are presented in Table 4.

Table 4. Reservoir and catchment information used in G-res.

Reservoir Information	
Country	Romania
Longitude of Dam	44◦26′7" N
Latitude of Dam	26◦6′10" E
Climate Zone	Temperate
Water Uses	Recreation, flood control, irrigation
Impoundment Year	1970
Reservoir Area	7.13 km^2
Reservoir Volume (Multiannual Average)	0.0185 km^3
Water Level (m above sea level)	70 masl
Maximum Depth	5 m
Mean Depth	4.2 m
Soil Carbon Content Under Impounded Area	0.8 kg C/m^2 [20]
Annual Wind Speed at 10 m	4 m/s
River Length Before Impoundment (m)	37.4 m
Reservoir Mean Global Horizontal Radiance	3.24 $kWh/m^2/d$ [31]
Mean Annual Air Temperature	13 °C [31]
Catchment Information	
Catchment Area	643 km^2
Population in the Catchment	3500000
Catchment annual runoff	700 mm/year [31]
Community wastewater treatment	Secondary
Industrial wastewater treatment	None
River Area Before Impoundment	0.2 km^2

3 Results and discussion

Based on the data provided, G-res computes not only the GHG footprint, but also some characteristic elements of the equivalent reservoir, as follows: Littoral area = 16.015%, Water residence time = 0.14 year and Annual discharge from Main Intake = 14.3 m^3/s.

The values of net GHG footprint for Colentina reservoirs, including the 95% confidence interval, are presented in Figure 4, while Figure 5 presents the net GHG emissions contributions for each equivalent reservoir water uses (services). The results are comparable to those reported in the literature [12, 20, 32].

The emission factors depend on the land type and use, and the data used in this simulation are presented in table 3. Under these assumptions, the landscape CO_2 emission rate for the Bucharest reservoirs for the pre-impoundment stage are 23 $gCO_2e/m^2/year$, which means 344 $tCO_2e/year$.

The values of CO_2 and CH_4 for the post-impoundment stage are:

- CO_2 release rate = 181 $gCO_2e/m^2/year$,
- CH_4 release rate = 121 $gCH_4/m^2/year$.

Thus, for post-impoundment stage, results a CO_2 emission rate for the Bucharest reservoirs of 302 $gCO_2e/m^2/year$.

Table 5 presents the CH_4 emissions from the reservoir associated with other anthropogenic sources. Given the fact that the water residence times is less than one year - very low, the CH_4 emissions associated with other anthropogenic sources is high and has an important share from the total CH_4 emission of post-impoundment.

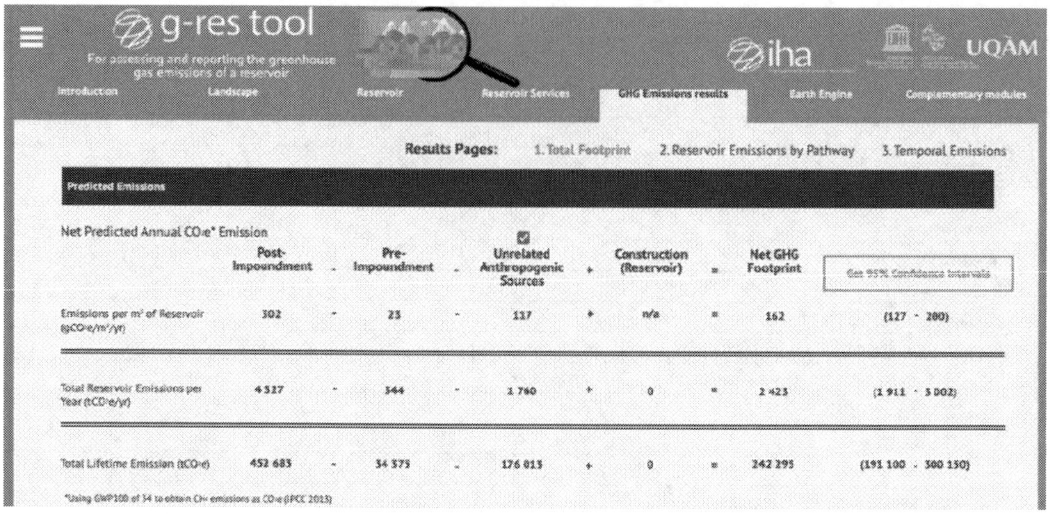

Fig. 4. Net GHG footprint for the equivalent reservoir (Colentina reservoirs).

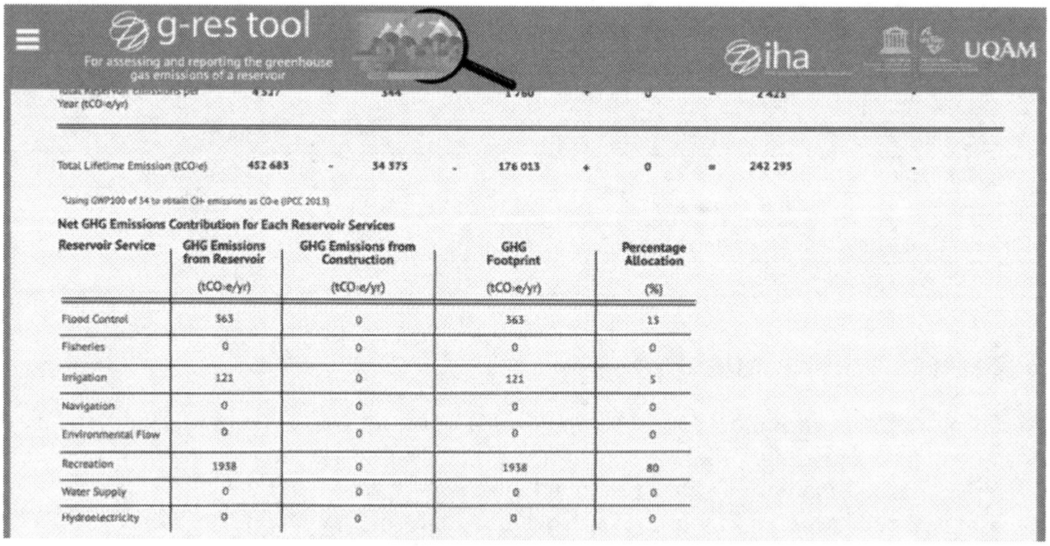

Fig. 5. Net GHG emissions contributions for each equivalent reservoir water uses (services).

Table 5. CH4 values associated to other anthropogenic sources.

Water Residence Time (year)	0.1
Reservoir CH_4 Emissions ($gCO_2e/m^2/year$)	120.8
Estimated CH_4 Release due to UAS ($gCO_2e/m^2/year$)	117.3
- due to land use ($gCO_2e/m^2/year$)	25.4
- due to sewage ($gCO_2e/m^2/year$)	91.9

Net reservoir footprint and CO_2 and CH_4 emission compared to reservoirs from same climate are presented in Figure 6 and Figure 7. As it can be seen in Figure 6, the Net reservoir footprint for Bucharest reservoirs, 162 $gCO_2e/m^2/year$ (Figure 4), is not very high compared to reservoirs from same climate. The small black arrow which indicates the position of the Bucharest reservoirs is placed on the beginning of the curve, and the values obtained for the reservoirs contained in the G-res Tool library range from 0 to over 3000 $gCO_2e/m^2/year$, the most frequent values being between 100 and 600 $gCO_2e/m^2/year$. Related to CH_4 post-impoundment emissions, no degassing is considered since the equivalent reservoir is shallow and the thermocline level is not known. As a result, only diffusive and bubbling mechanisms are considered, representing 85% and 15% respectively.

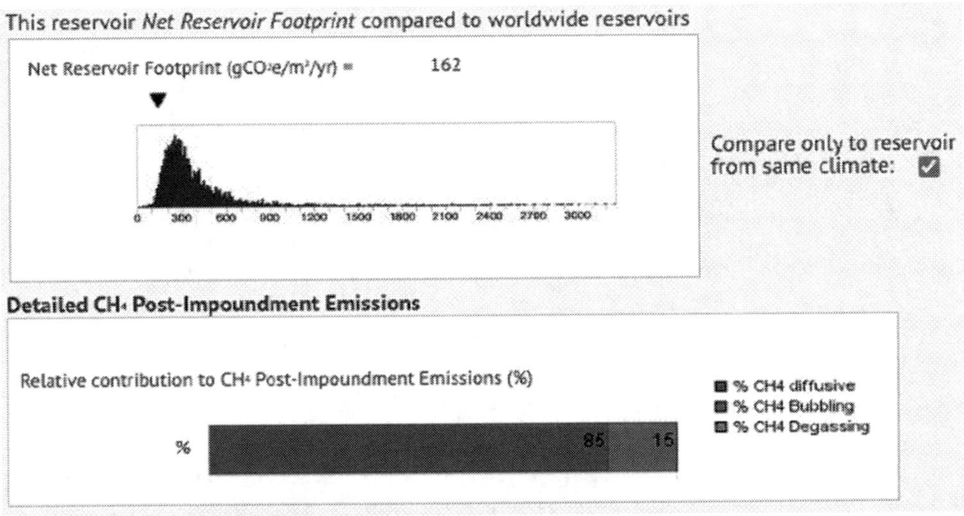

Fig. 6. Net reservoir footprint of the equivalent reservoir compared to reservoirs from same climate and detailed contribution to CH_4 post-impoundment emissions.

Figure 7 presents in detail CH_4 and CO_2 emissions for Bucharest reservoirs in contrast with reservoirs from same climate. In terms of CH_4 diffusive emissions, the reservoir is placed in the first third of the descending branch/curve, while the bubbling emissions is in the high values area. CO_2 diffusive emissions place the reservoir in the first third of the descending branch/curve, as well.

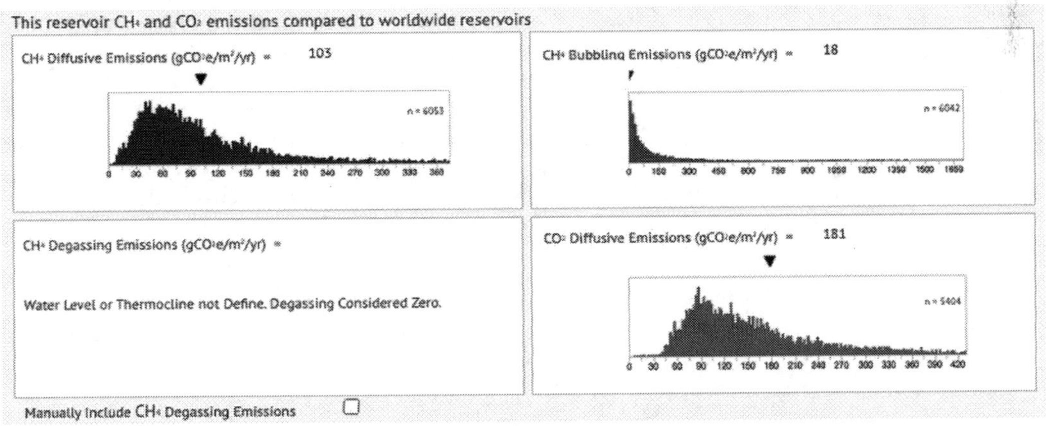

Fig. 7. CH_4 and CO_2 emissions of the equivalent reservoir compared to reservoirs from same climate.

Finally, the temporal variation of GHG by emission pathways of the analysed equivalent reservoir is presented in Figure 8.

According to the simulation values, GHG intensity exponentially decreased (Figure 8), at this date (50th year) representing just approximately 11% (~200 gCO_2e/m^2/year) from the initial value (the one from the completion, just after the damming of the water course, i.e. 1970).

 https://doi.org/10.1051/e3sconf/202340402001

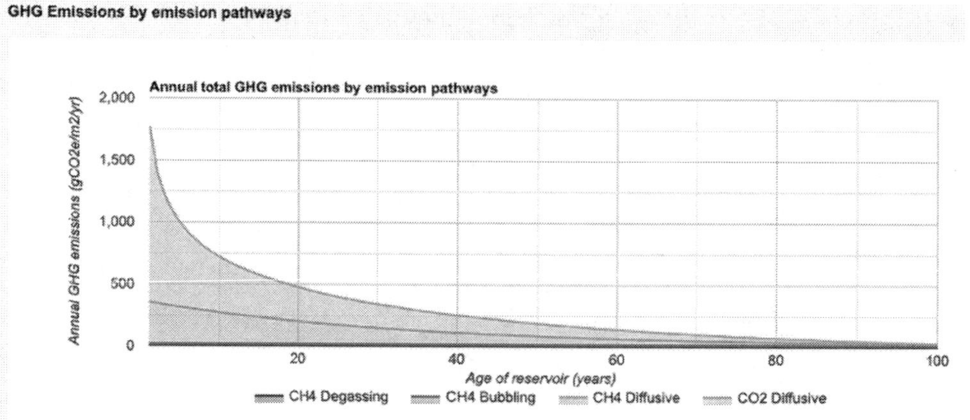

Fig. 8. Temporal variation of GHG by emission pathways.

4 Conclusions

This paper determined carbon footprint related to the 10 reservoirs on Colentina river in Bucharest, considered in the analysis by means of an equivalent reservoir, using GHG Reservoir Tool (G-res).

There were presented entry data, hypotheses, and the use of G-res tool, applied for determining GHG emissions of the equivalent reservoir. Thus, we now have a good picture of the 10 reservoirs on Colentina river contribution to the overall GHG emissions in Bucharest.

According to these results we can appreciate that, at this moment, the 10 reservoirs on Colentina river in Bucharest do not have a substantial contribution to GHG emissions, and in the next 25-30 years they will reach emissions comparable to those obtained for the undeveloped version of the Colentina river. As a result, in our opinion, these reservoirs represent a plus both in terms of providing adequate recreational spaces for the population, and increased benefits to the microclimate and population of Bucharest.

References

1. IPCC (Intergovernmental Panel on Climate Change), "2006 IPCC Guidelines for National Greenhouse Gas Inventories", edited by S. Eggleston, L. Buendia, K. Miwa, T. Ngara and K. Tanabe (Published by the Institute for Global Environmental Strategies (IGES)), Hayama, Japan on behalf of the IPCC (2006)
2. IEA Hydropower, "Guidelines for quantitative analysis of net GHG emissions from reservoirs" (2012)
3. World Bank, "Greenhouse Gases from Reservoirs Caused by Biogeochemical Processes", World Bank, Washington, DC, USA (2017)
4. J. J. Cole, Y. T. Prairie, N. F. Caraco et al., Plumbing the global carbon cycle: Integrating inland waters into the terrestrial carbon budget, Ecosystems **10**, 172–185 (2007)
5. N. Barros, J. J. Cole, L. J. Tranvik, Y. T. Prairie, D. Bastviken, V. L. M. Huszar, P. del Giorgio and F. Roland, Carbon emission from hydroelectric reservoirs linked to reservoir age and latitude, Nature Geosci. **4**, 593–596 (2011)
6. D. Butman and P. A. Raymond, Significant efflux of carbon dioxide from streams and rivers in the United States, Nature Geosci. **4**, 839–842 (2011)

7. P. A Raymond, J. Hartmann, R. Lauerwald et al., Global carbon dioxide emissions from inland waters, Nature **503**, 355–359 (2013)
8. B. R. Deemer, J. A. Harrison, S. Li, J. J. Beaulieu, T. DelSontro, N. Barros et al., Greenhouse gas emissions from reservoir water surfaces: A new global synthesis, BioScience **66**, 949–964 (2016)
9. T. DelSontro, J. J. Beaulieu and J. A. Downing, Greenhouse gas emissions from lakes and impoundments: Upscaling in the face of global change, Limnology and Oceanography Letters **3**(3), 64–75 (2018)
10. B. R. Deemer and M. A. Holgerson, Drivers of methane flux differ between lakes and reservoirs, complicating global upscaling efforts, JGR Biogeosciences **126** (2021)
11. J. A. Rosentreter, A. V. Borges, B. R. Deemer et al., Half of global methane emissions come from highly variable aquatic ecosystem sources. Nat. Geosci. **14**, 225–230 (2021)
12. R. Gruca-Rokosz, Quantitative fluxes of the greenhouse gases CH_4 and CO_2 from the surfaces of selected polish reservoirs, Atmosphere **11** (2020)
13. Y. T. Prairie, J. Alm, J. Beaulieu et al., Greenhouse gas emissions from freshwater reservoirs: What does the atmosphere see?, Ecosystems **21**, 1058–1071 (2018)
14. Y. T. Prairie, J. Alm, A. Harby, S. Mercier-Blais, R. Nahas, "The GHG Reservoir Tool (G-res) Technical documentation", Updated version 3.0 (2021-10-27), UNESCO/IHA research project on the GHG status of freshwater reservoirs. Joint publication of the UNESCO Chair in Global Environmental Change and the International Hydropower Association 73 pages (2017)
15. IHA (International Hydropower Association), Carbon Emissions from Hydropower Reservoirs: Facts and Myths (2021)
16. L. Scherer and S. Pfister, Hydropower's biogenic carbon footprint, Plos one **11**(9) (2016)
17. T. Vilela and J. Reid, Improving hydropower choices via an online and open access tool, Plos one **12**(6) (2017)
18. X. Hao, Y. Ruihong, Z. Zhuangzhuang et al., Greenhouse gas emissions from the water–air interface of a grassland river: a case study of the Xilin River, Sci. Rep. **11,** 2659 (2021)
19. E. León-Palmero et al, Environ. Res. Lett. **15** (2020)
20. I. V. Ion and A. Ene, Evaluation of greenhouse gas emissions from reservoirs: A review, Sustainability **13**, 11621 (2021)
21. S. H. Ortega, C. R. González-Quijano, P. Casper, G. A. Singer, M. O. Gessner, Methane emissions from contrasting urban freshwaters: Rates, drivers, and a whole-city footprint, Glob Change Biol. **25,** 4234– 4243 (2019)
22. https://sites.google.com/site/municipiulb/matrix
23. RoWater, "Lucrări hidrotehnice în exploatare", https://arges-vedea.rowater.ro/wp-content/uploads/2021/02/Lucrari-hidrotehnice-in-exploatare.pdf
24. Wikipedia, "Râul Colentina", https://ro.wikipedia.org/wiki/R%C3%A2ul_Colentina
25. Aquaproiect, "Atlasul Cadastrului Apelor din România", Ministerul Mediului, Bucureşti, Romania (1992)
26. O. Cocoş, "Managementul apei în Municipiul Bucureşti (First edition)", Publisher Ars Docendi, Bucharest, Romania (2006).

27. www.hydrop.pub.ro/attachments/article/138/2.%20Dorin%20Pavel%20Dir%20tehn%201934.pdf
28. RoWater, "The Project Updated Management Plan of The Hydrographic Space Argeş-Vedea" (2021)
29. RoWater, "The plan of Flood risk management Argeş - Vedea Water Basin Administration", National Administration "Apele Romane", Bucharest, Romania, (2017)
30. https://buletin.de/bucuresti/raport-de-mediu-in-bucuresti-sunt-mai-multi-oameni-decat-copaci
31. MeteoBlue, www.meteoblue.com
32. T. Delsontro, D. F. Mcginnis, S. Sobek, I. Ostrovsky, B. Wehrli, Extreme methane emission from Swiss hydropower reservoir: Contribution from bubbling sediments, Environ. Sci. Technol. **44**, 2419–2425 (2010)

Optimization of the blade profile of a vertical axis wind turbine based on aerodynamic analysis

Alina Fazylova[1,2*], *Michail* Malamatoudis[3], and *Panagiotis* Kogias[3]

[1]Satbayev University, Department of Robotocs and engineering tools of automation, 050013, Almaty, Kazakhstan
[2]Almaty University of Power Engineering and Telecommunications named after G.Daukeev, Department Electronics and Robotics, 050013, Almaty, Kazakhstan
[3]International Hellenic University, Department of Physics, University Campus II, Agios Loukas, 65403, Kavala, Greece

Abstract. This paper explores the influence of the angle of attack on the aerodynamic characteristics of the blade profile. The paper presents calculations, modeling and graphical representation of the blade shape during rotation around the axis. Using the given parameters, such as the length of the blade, the radius of the upper and lower boundaries, as well as the angle of inclination of the blade, the calculation of the coordinates of the points of the blade profile for various angles of rotation is given. The cross-sectional area, volume and mass of the blade were also calculated. Appropriate calculations were made to approximate the center of mass of the blade. To evaluate the influence of the angle of attack on the blade profile, the angles of attack were calculated for various angles of rotation.

1 Introduction

The study of the aerodynamic properties of the blades is an important task in the development of wind power plants, [1]. Understanding the shape and characteristics of the blades as they rotate allows you to optimize the efficiency and reliability of power plants, as well as reduce the negative impact on the environment [2]. To demonstrate this approach, an example with blades of certain parameters was considered in the work: blade length (L) 4 meters, upper boundary radius (R_top) 2 meters, lower boundary radius (R_bottom) 1 meter and blade inclination angle (α) 15 degrees. This work can be considered in the context of other existing studies, which are also devoted to the study of the shape of the blades and their aerodynamic properties. For example, in the study [3], an analysis was made of the influence of various parameters of the blades on their efficiency and the stability of the operation of wind power plants. The results of this study confirmed the importance of the optimal blade shape for achieving high efficiency and sustainability of wind power plants. Another study carried out in [4] focused on the aerodynamic design of blades using a computer simulation method. The authors of the study used modern

* Corresponding author: sharf_06@inbox.ru

software tools to optimize the shape of the blades in order to minimize drag and maximize lift. The results of this study proved that optimized blades can increase the efficiency and power output of wind turbines. Another interesting study, described in the article [5], studied the influence of the aerodynamic characteristics of the blades on the noise emission of wind turbines. The authors conducted experiments and simulations to investigate the relationship between the shape of the blades, their aerodynamic properties and the level of noise generated during the operation of wind power plants. The results indicate that certain changes in the shape of the blades can reduce noise emissions, which is an important aspect for improving the environmental suitability of wind power plants. Thus, these research examples demonstrate the importance of studying the shape and aerodynamic properties of wind turbine blades. Understanding these factors allows you to optimize the operation of plants, increasing their efficiency, sustainability and environmental suitability.

2 Materials and methods

In this paper, the following blade parameters of a vertical-axis wind turbine were used [6]: blade length (L): 4 meters, upper boundary radius (Rtop): 2 meters, lower boundary radius (R_bottom): 1 meter, blade inclination angle (α): 15 degrees. These parameters were set and used for calculations.

To understand the shape and aerodynamic properties of a vertical-axis wind turbine blade, the blade characteristics are calculated and analyzed. In particular, the following parameters are calculated:

- chord length (C) - the length of a straight line connecting the upper and lower boundaries of the blade. This parameter helps to determine the scale of the blade and its geometric properties.

– blade thickness (T) – the distance between the upper and lower boundaries of the blade. This parameter affects the aerodynamic characteristics of the blade and may be important in design.

– coordinates of blade profile points – calculation of coordinates (x, y) of points on the blade profile at different angles of rotation (θ). This allows visualization of the blade shape and aerodynamic analysis such as angle of attack.

– blade cross-sectional area – the area covered by the blade in the transverse direction. This parameter is important for determining the geometric properties and characteristics of the blade.

– blade volume – the volume occupied by the blade in three-dimensional space. The calculation of the volume allows you to estimate the mass and density of the blade.

– blade mass – mass of the blade, determined on the basis of its volume and density of the material. This parameter is important in the analysis of the dynamics and mechanical properties of the blade.

– center of mass of the blade – calculation of the approximate coordinates of the center of mass of the blade. The center of mass is an important parameter when analyzing the static and dynamic behavior of a blade.

The chord length (C) is calculated using the given parameters and mathematical operations. The following equation was used to calculate the chord length:

$$C = 2 \cdot (R\_top - R\_bottom) \cdot tg\ \alpha. \quad (1)$$

The blade thickness (*T*) is calculated from the value of C using the corresponding formula, [7]:

$$T = 0{,}18 \cdot C. \quad (2)$$

The (x, y) coordinates of the blade profile points are calculated using the given parameters and rotation angles θ. The coefficients in the equation for calculating the coordinates (x, y) of the blade profile points were obtained from the NACA (National Advisory Committee for Aeronautics) airfoil. NACA airfoils are a system of standardized airfoils developed by the US National Air Research Committee (NACA). These profiles have a mathematical description that allows you to calculate the coordinates of points on their surface. This equation uses the formula for calculating the NACA 4 profile [8, 9]. Thus, the following equations were obtained for calculating the coordinates:

$$x = \frac{L}{2}(1 - \cos\theta);$$
$$y = R_b + [R_{top} - R_{bottom}] \cdot \sin\theta + T \cdot 10{,}2969 \cdot \sqrt{\frac{x}{c}} - 0{,}126 \cdot \frac{x}{C} - 0.3516 \cdot \left(\frac{x}{C}\right)^2 + 0{,}2843 \cdot \left(\frac{x}{C}\right)^3 - 0{,}1036 \cdot \left(\frac{x}{C}\right)^4. \quad (3)$$

To calculate the cross-sectional area of the blade (S), the following equation was used:

$$S = L \cdot T. \quad (4)$$

The volume (V) and mass of the blade (m) are calculated based on the cross-sectional area, blade length and material density. In this work, the following equations were used:

$$V = S \cdot L; \quad (5)$$

$$m = L \cdot S \cdot \rho, \quad (6)$$

where ρ, is material density.

To calculate the approximate center of mass, the methods of summing the x and y coordinates of all points of the blade profile were used and the midpoint of the blade profile was calculated by dividing the sum of coordinates by the total number of profile points [10]. To calculate the approximate value of the moment of inertia, the blade was presented as a rectangular rod with a given blade width (W) and blade material density. The angle of attack is calculated according to the following formula, [11]:

$$\alpha = arctg\left(\frac{y - y_c}{x - x_c}\right), \quad (7)$$

where: α - angle of attack; y - vertical coordinate of a point on the blade; x - horizontal coordinate of a point on the blade; y_c - y-coordinate of the center of mass of the blade; x_c - x-coordinate of the center of gravity of the blade.

To calculate the midpoint of the blade profile, you can use the formulas:

$$x_c = \frac{\Sigma x}{N}; \quad (8)$$

$$y_C = \frac{\Sigma y}{N}, \quad (9)$$

where N is the total number of blade profile points (in this case, $N = 12$).

To calculate the approximate value of the moment of inertia of the blade, you can use the formula for the moment of inertia of a rectangular rod about its axis of rotation. Let's assume that the blade is a rectangular rod with the following parameters: blade width (W) = 0.1 meters, blade material density (ρ) = 1.2 kg/m³ (approximate value). The moment of inertia of a rectangular rod about its axis passing through the center of mass can be calculated using the formula [12]:

$$I = \frac{1}{12} \cdot m \cdot (w^2 + T^2). \tag{10}$$

3 Results

In accordance with formula (1), the chord length is:

$$C = 2 \cdot (2-1) \cdot tg\,15 = 0{,}342\ m. \tag{11}$$

Now, using the value of C, the blade thickness T has been calculated according to expression 2:

$$T = 0{,}18 \cdot 0.342 = 0{,}0616\ m. \tag{12}$$

Next, you need to determine the volume of the blade and its mass. To calculate the mass of the blade, the density of the material (ρ) is required, which is equal to 1.2 kg/m³ in accordance with expressions 5 and 6:

$$V = 0{,}2464 \cdot 4 = 0{,}9856\ m^3. \tag{13}$$

$$m = 0{,}2464 \cdot 4 \cdot 1{,}2 = 1{,}181 kg. \tag{14}$$

Next, the approximate center of mass of the blade was calculated. For this, the sum of all x and y coordinates was calculated and the midpoint of the blade profile was calculated:

$$\Sigma x = 0 + 0.342 + 1.268 + 2.000 + 2.732 + 3.658 + 4.000 + 3.658 + 2.732 + 2.000 + 1.268 + 0.342 = 24.320; \tag{15}$$

$$\Sigma y = 1 + 1.0502 + 1.2771 + 1.1839 + 0.9352 + 0.6694 + 0.5716 + 0.6694 + 0.9352 + 1.1839 + 1.2771 + 1.0502 = 11.9624. \tag{16}$$

In accordance with formulas 8 and 9, the approximate coordinates of the center of mass of the blade are:

$$x_c = \frac{24{,}320}{12} = 2{,}0267 m; \tag{17}$$

$$y_c = \frac{11{,}9624}{12} = 0{,}9969 m. \tag{18}$$

Using formula 3, the coordinates of the blade profile points were calculated for various angles of rotation θ. The rotation angle θ varies from 0 to 360 degrees in 30 degree increments. As a result of calculations, the following table was obtained with coordinates (x, y) (Table 1):

Table 1. Coordinate calculation.

Rotation angle θ	**x (meters)**	**y (meters)**
0	0	1
30	0.342	1.0502
60	1.268	1.2771
90	2.000	1.1839
120	2.732	0.9352
150	0.342	1.0502
180	3.658	0.6694
210	4.000	0.5716
240	3.658	0.6694
270	2.732	0.9352
300	2.000	1.1839
330	1.268	1.2771
360	0.342	1.0502

Thus, the coordinates (x, y) were obtained for each value of the rotation angle θ in the range from 0 to 360 degrees with a step of 30 degrees. These coordinates represent the blade profile shape with given parameters such as blade length, upper and lower boundary radius, and blade pitch angle. Graphically represented coordinates (x, y) allow you to visually see the shape of the blade as it rotates around the axis. This information can be useful in designing and analyzing the aerodynamic properties of a blade, such as the forces created by the wind as it spins. To obtain a blade shape graph, a program code was written in Python:

```
pip install numpy
pip install matplotlib
import numpy as np
import matplotlib.pyplot as plt
# Specified parameters
L = 4
R_top = 2
R_bottom = 1
alpha = 15
# Calculation of chord length C
C = 2 * (R_top - R_bottom) * np.tan(np.radians(alpha))
# Calculation of blade thickness T
T = 0.18 * C
# Calculation of rotation angle values theta
theta = np.arange(0, 361, 30)
# Calculate coordinates (x, y)
x = (L / 2) * (1 - np.cos(np.radians(theta)))
y = R_bottom + (R_top - R_bottom) * np.sin(np.radians(theta)) + T * (
   0.2969 * np.sqrt(x / C) - 0.126 * (x / C) - 0.3516 * (x / C) ** 2 +
   0.2843 * (x / C) ** 3 - 0.1036 * (x / C) ** 4
)
# Graphical representation
plt.plot(x, y)
plt.xlabel('x (in meters)')
plt.ylabel('y (in meters)')
plt.title('Shape of the blade during rotation')
plt.grid(True)
plt.show()
```

After running the program code, the blade shape shown in Figure 1 was obtained.

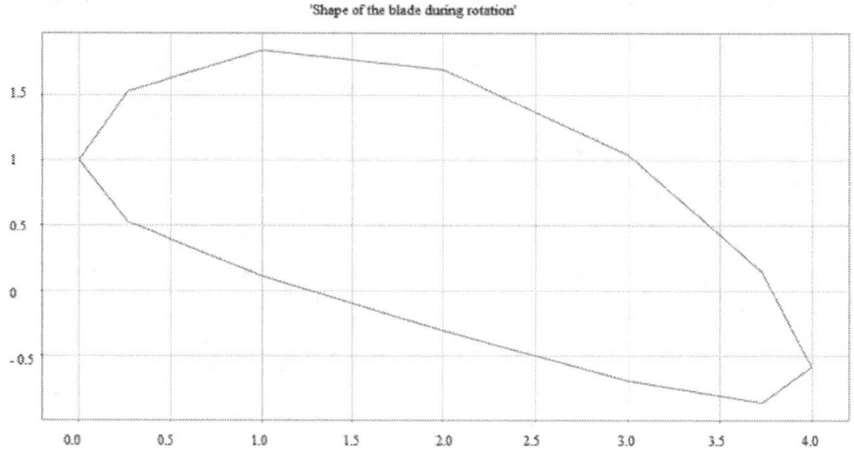

Fig. 1. Blade shape.

The Figure 1 shows the blade shape of a vertical axis wind turbine during rotation. The blade has a curved shape, reminiscent of the profile of an aircraft wing. It consists of two curves - upper and lower surfaces. The dimensions of the blade are determined by the parameters specified in the code, such as the length of the blade, the top and bottom radii. The angle of attack specified in the code determines the inclination of the blade relative to the air flow. The blade profile shape is calculated using the 4-series NACA equation. The graph allows you to visualize the shape of the blade as it rotates around a vertical axis. In accordance with formula 10, the moment of inertia was calculated:

$$I = \frac{1}{12} \cdot 1{,}181 \cdot (0{,}1^2 + 0{,}0616^2) = 0.00140115\ kg \cdot m^2 \tag{19}$$

Further, according to equation 7, the angle of attack was determined and the following values were obtained:

Angle of attack at 30 degrees: α_30 ≈ 0.79 degrees
Angle of attack at 60 degrees: α_60 ≈ 6.70 degrees
Angle of attack at 90 degrees: α_90 ≈ 16.70 degrees
Angle of attack at 120 degrees: α_120 ≈ 25.34 degrees
Angle of attack at 150 degrees: α_150 ≈ 33.20 degrees
Angle of attack at 180 degrees: α_180 ≈ 41.19 degrees
Angle of attack at 210 degrees: α_210 ≈ 49.48 degrees
Angle of attack at 240 degrees: α_240 ≈ 58.34 degrees
Angle of attack at 270 degrees: α_270 ≈ 68.34 degrees
Angle of attack at 300 degrees: α_300 ≈ 80.57 degrees
Angle of attack at 330 degrees: α_330 ≈ 101.13 degrees

Thus, the calculation was performed and the coordinates of the points of the blade profile for various angles of rotation θ in the range from 0 to 360 degrees with a step of 30 degrees were presented. The calculations were carried out using given parameters, such as the blade length, the radius of the upper and lower boundaries, and the angle of the blade. Coordinates (x, y) represent the shape of the blade profile as it rotates around the axis. The graphical representation of these coordinates clearly demonstrates the shape of the blade in space. The information obtained can be useful in carrying out subsequent studies in the design and analysis of the aerodynamic properties of the blade. For example, the resulting shape of the blade profile allows you to evaluate its aerodynamic characteristics, such as the air flow around the profile, the angle of attack: The calculated values of the angle of attack allow you to estimate at what angle the blade meets the air flow. The angle of attack is a key parameter for aerodynamic analysis, as it determines the lift and drag forces, flow dynamics and the shape of the blade flow. The aerodynamic properties of the blade are directly related to the forces generated. Based on the data provided, it is possible to estimate the magnitude of the lift (air support) and drag forces for each value of the angle of rotation θ. This can be useful in determining the efficiency of a blade in handling wind and ensuring stability. The estimated moment of inertia allows you to evaluate the inertial properties of the blade. The moment of inertia can be an important parameter in the analysis of the dynamics of the rotation of the blade, for example, in the case of wind power installations.

4 Conclusion

This article explores the influence of the angle of attack on the aerodynamic characteristics of the blade profile. The paper presents calculations, modeling and graphical representation of the blade shape during rotation around the axis. Using the given parameters, such as the length of the blade, the radius of the upper and lower boundaries, as well as the angle of inclination of the blade, the coordinates of the points of the blade profile for various angles of rotation were calculated. The cross-sectional area, volume and mass of the blade were also calculated. Appropriate calculations were made to approximate the center of mass of the blade. The results of calculations and modeling made it possible to obtain the shape of the blade profile at various angles of rotation (θ) from 0 to 360 degrees with a step of 30 degrees. The coordinates (x, y) of the blade profile points were presented in a table and graphically displayed on the graph.

These results may be useful for further design and analysis of the aerodynamic properties of wind turbine blades. This work is of practical importance in the design and analysis of the aerodynamic characteristics of the blades, such as the forces created by the wind during rotation. The results obtained can be used to optimize the shape and improve the efficiency of wind turbines.

References

1. V. Khudoyarov et al. J. Phys.: Conf. Ser., **1679**, 052078 (2020)
2. C. Genç, A. Sakalli, I. Stoyanov, T. Iliev, G. Mihaylov, I. Beloev, E3S Web of Conferences, **207**, 02013 (2020)
3. C. Genç, A. Sakalli, I. Stoyanov, T. Iliev, G. Mihaylov, I. Beloev, E3S Web of Conferences, **207**, 02014 (2020)
4. M. Zhou, M. Wang, J. Li and G. Li, CSEE Journal of Power and Energy Systems, **3**(1), pp. 74-83 (2017)
5. A. Fazylova, G. Balbayev, D. Ilieva, M. Aliyarova, E3S Web of Conferences, **180**, 02001 (2020)
6. A. M. Chowdhury, H. Akimoto, Y. Hara, Renewable Energy, **85**, pp.327-337, (2016)
7. N.N. Barbashov, et al. IOP Conf. Ser.: Mater. Sci. Eng., **862,** 032047 (2020)
8. A. Fazylova, T. Iliev, *A New Type of Wind Generator Blades Mechanical Automatic Control System*, in Proceedings of the 2022 International Conference on Communications, Information, Electronic and Energy Systems (CIEES), Veliko Tarnovo, Bulgaria, (2022)

9. D. Daskalaki, J. Fantidis, P. Kogias, The Journal of CIEES, **1**(1), pp. 23–28, (2021) https://doi.org/10.48149/jciees.2021.1.1.4
10. A. Terziev, Y. Panteleev, M. Ivanov, IOP Conf. Ser.: Mater. Sci. Eng., **1032** 012041, (2021), https://doi.org/10.1088/1757-899X/1032/1/012041
11. L.R. Abdullina et al., IOP Conf. Ser.: Earth Environ. Sci., **981**, 032090 (2022)
12. J.H. Ferziger, M. Peric, Computational Methods for Fluid Dynamics. (Springer Berlin, Heidelberg, 2012)

Detailed infrared diagnostics of photovoltaic panels for higher safety and optimal energy generation

Ventsislav Keseev*

University of Ruse "Angel Kanchev", Department of Telecommunications, 8 Studentska str., 7017 Ruse, Bulgaria

Abstract. The Photovoltaic (PV) systems are one of the key renewable energy sources that are becoming increasingly popular, but they still have many drawbacks compared to conventional energy sources. Their main disadvantages are that they require a lot of installation space and they have low investment return ration. Both could be overcome by local installation of the PV plants as close as possible to the end consumer, and the best case is the PV panels to be installed on the buildings, but this could lead to higher fire risk for the residents. Different PV panels degrade with different speeds and work under slightly different conditions. These problems require at least periodic PV panel diagnostics and the thermal infrared (IR) inspection seems to be the best solution. The prerequisites for correct IR diagnostics are presented. One of the requirements for detailed IR inspection is the thermal camera to capture each PV cell with at least 5 by 5 pixels in horizontal and vertical directions. A methodology for pixel size calculation and IR monitoring system design is developed and presented in this work.

1 Introduction

The world is facing global environmental and climate changes, and the renewable energy sources could be a good solution. The photovoltaic (PV) systems have the potential to solve these problems, but they still have many drawbacks. They are not appropriate for every climate and different solutions for their effectiveness improvement are needed in this regard because the meteorological parameters as wind, solar radiation, air temperature, air humidity, atmospheric pressure and precipitation differ, [1]. The problem is that the I-V curve of the PV panels change together with the ambient conditions as temperature, angle of illumination, intensity and spectrum of radiation, and with the array degradation too, [2]. At the same time, the PV systems have a low investment return ratio compared to other conventional energy sources and it could be even worst because studies conclude that in certain cases there are significant amounts of unused energy, which further seriously reduces their efficiency, [3,4]. In such cases, it is important to increase the energy utilization if the PV investment is to be optimal.

* Corresponding author: vkeseev@uni-ruse.bg

There are many ways for PV system optimization. One option is the PV system to be combined with another energy generation system. In this regard, a study investigated the optimum tilt angle of a photovoltaic-wind hybrid system which utilizes both wind and solar energy by attaching the PV modules to the blades of a vertical-axis wind turbine, [5]. Another option is to optimize the operation of a photovoltaic system by cogeneration, simultaneous production of electricity and low-temperature water heating, and increasing efficiency by achieving a balance between electricity and heat production, [6]. Most of the available optimization methods are still experimental and often not applicable, and they don't solve certain hazards related to PV system utilization.

From another point of view, the PV systems require a lot of installation space. The problem is that the human population grows and the land is needed for food production.

The mentioned drawbacks could be solved with local installation of PV infrastructure on residential buildings. PV systems with small power are an efficient and sustainable method of generating electricity. Mounted on roofs, facades, sheds or other free spaces, these systems are currently the most cost-effective option for utilizing free areas, reducing electricity costs and sustainable production of electricity, without additional losses from its transmission, [7]. One of the main drawbacks is that this solution leads to higher fire risks and requires at least periodic PV panel diagnostics. However, the rooftop solar photovoltaic potential has been estimated in many countries, such as Japan, India, Spain, and Switzerland using various methods, [8]. In this regard, the infrared (IR) cameras are becoming increasingly popular for thermal inspection of PV panels, necessary for timely discovery of defects for safety purposes and for overall system optimization. A study evaluated the mid-life degradation of PV plant and 52.5% of the 360 PV modules showed certain defects. The conclusion is that in a PV plant, the degraded defective PV modules must be replaced immediately because of mismatch losses of strings and risk of fire, [9]. At the same time, a study compared and analysed the performance and degradation differences of two photovoltaic modules from different producers after 15 years of field operation and concluded that they are with very different resulting observed changes in their operational and physical characteristics, [10]. Another investigation concluded that the load-bearing structures and the photovoltaic panels must be able to withstand mechanical loads both from their own weight and from snow and wind [11]. The more extreme weather conditions lately, as strong wind and heil, can cause sudden damage to PV panels and to increase considerably the fire risk.

The quality differences between PV panels of different producers and the chances for unexpected damage require periodic IR inspections to be done in all the phases of their life, even at the time of first installation for warranty purposes. In this regard, the creation of virtual laboratories for remote investigations of photovoltaic modules could also help a lot because they allow operating the lab independently of the weather conditions and the time of the day, [12].

The thermal IR diagnostic of PV panels is one of the most widespread non-invasive technique used nowadays. It has many advantages and could be used on spot, but in order the inspection to be reliable it must be done under certain conditions. The solar irradiance should be over 600W/m2, [13]. Another research concludes that the irradiance must be over 700 W/m2 and that the measurements are best to be done in cold winter day, [14]. The camera should be looking at the panels from a position with certain horizontal and vertical angles. The ideal horizontal angle of thermographic inspection must be the angle which minimizes the solar reflection and does not increase the reflectivity and it depends on the panel orientation, and time of the year, [15]. The vertical angle between the module plane and the camera should be between 30 and 90 degrees, [16]. For detailed IR inspection the camera must capture each solar cell with standard sizes of 16 by 16 cm with at least 5 by 5 pixels horizontally and vertically, [13]. This is the minimum required image resolution to detect certain smaller defects. More pixels means a more accurate diagnostic process, but more costly too, because it requires either more expensive higher resolution cameras, or more cameras, or more pictures to be taken with one moving one. Usually, one IR camera must capture more than one PV panels and the distance from the sensor to each separate PV cell is different. At the same time the viewing angles for each solar cell are also slightly different. This leads to different pixel sizes for each solar cell and hence to different number of capturing pixels. It is important the furthest from the camera solar cell to be captured with at least 5 by 5 pixels because it also should be diagnosed in detail. If this is achieved, it means that all other solar cells will be captured with more than the minimum required resolution.

Most of the prerequisites are matter of right camera position, time of the day and season. The one requiring over 5 pixels per solar cell in each direction poses certain requirements towards the IR camera and its installation place, but no any precise methodologies for calculating the pixel size and for choosing the right camera are found.

The aim of this work is to present developed methodology for pixel size calculation, for correct selection of camera sensor resolution and for finding the right installation distance. Such methodologies are needed for experimental PV system designs, as well as for education of installers of such systems because people working in the field usually have electrical or electronic background and no experience with imaging techniques.

2 Methodology

The sizing methodology is presented in a structured way in two main steps. The concept is visualized through comprehensible drawings. The first logical step is defining the minimum distance from certain IR camera with certain viewing angles to the PV panels because first of all it must be able to capture certain number of them. The second logical step is the calculation of the pixel size and pixel number per PV cell for certain IR sensor resolution and the selection of the right thermal camera based on this criteria.

For the monitoring system sizing, the minimum distance from certain IR camera to the PV panels should be calculated first, based on its viewing angles and the size of the monitored PV installation. The stationary camera should be as close as possible to the monitored PV panels for more precise diagnostics, but it should be able to capture all of them. If the camera is with the more expensive pan and tilt system, then this requirement is not fully valid. The viewing angles of different IR cameras are usually within the range of certain common values and the right one with the right installation distance should be found.

The second step could vary. One way is the pixel size to the most distant PV panels to be calculated for certain camera with certain standard resolution. The other more efficient way is the needed IR sensor resolution to be calculated based on the distance between the thermal camera and the furthest PV panel. The calculations depend on the thermal camera parameters, on its vertical and horizontal viewing angles, on the PV panel cell sizes and on the distance between them and the camera. They are usually done for the worst-case scenario, which means that the distance to the furthest monitored PV cells should be used for the calculations and the selection of the needed camera sensor resolution. This requirement usually comes from the fact, that even the furthest PV cells need to be monitored in certain details. If a camera with such viewing angles and sufficient resolution doesn't exist or it is too expensive, then two cameras could be installed or one with pan and tilt system. This methodology is for stationary mounted IR cameras, but it could be used and for a monitoring system design with pan and tilt ones.

3 Defining the minimum installation distance from certain IR camera with certain viewing angles to the PV panels

The minimum distance from the IR camera to the PV panels should be calculated first, based on its viewing angles and the size of the monitored PV installation. This requires the width of the closest first row of PV panels to be measured first. The worst-case scenario is when the camera is installed against the middle of the first row (centred to the first row) because then its sensor must be with the highest horizontal viewing angle. In this case the horizontal viewing plane is consisted by 2 right angle triangles, Figure 1. Usually, the horizontal viewing angle must be wider than the vertical, so it must be calculated first, and then it usually corresponds to certain comparatively standard vertical one, which is wide enough. In this case, for each right angle triangle is valid equation (1), where "w" is the width of the closest first row of PV panels, "d" is the distance between the camera and the PV panels and "α" is the horizontal IR camera viewing angle.

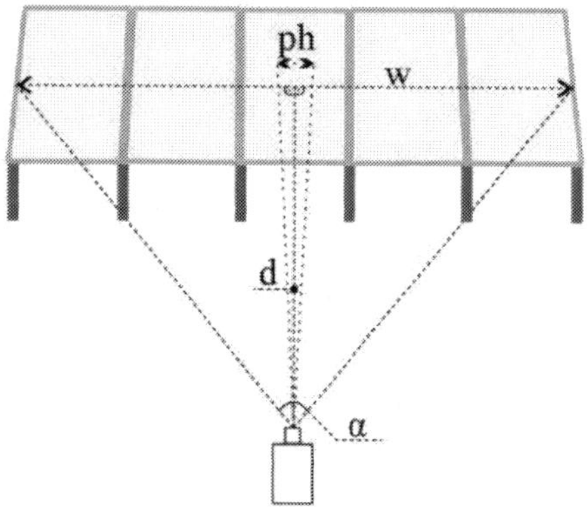

Fig. 1. Horizontal field of view of the thermal camera to the photovoltaic panels.

$$tg\left(\alpha/_2\right) = \frac{\frac{w}{2}}{d} = \frac{w}{2d} \tag{1}$$

If the camera horizontal viewing angle and the width of the first PV panel row are known, then the needed installation distance could be calculated by equation (2):

$$d \geq \frac{w}{2tg\left(\alpha/_2\right)} \tag{2}$$

If the installation distance and the width of the first PV panel row are known, then the minimum horizontal IR sensor viewing angle α could be calculated by equation (3):

$$\alpha \geq 2arctg\left(\frac{w}{2d}\right) \tag{3}$$

The installation distance could be either directly measured with the help of laser tape measure or calculated from the measured horizontal distance "l" and vertical installation height "h" of the camera related to the installation plane of the PV panels by (4), Figure 2:

$$d = \sqrt{h^2 + l^2} \quad (4)$$

There are IR camera sensors with 45° horizontal wide-viewing angles on the market and their vertical viewing angles are usually less than that. For example, the Mobotix thermal camera has 45° horizontal and 32° vertical image angles. Table 1 presents the results from a search for most common standard resolutions and available viewing angles (field of view) for modern IR cameras. It is visible that for the medium and professional class ones the available sensor viewing angles are about the same. For lower class IR cameras with resolution of 160 x 120 there are sensors with wider field of view, but their low resolution is usually enough for single PV panel diagnostic. In all examples the horizontal viewing angles are wider than the vertical ones because in most cases this is the need.

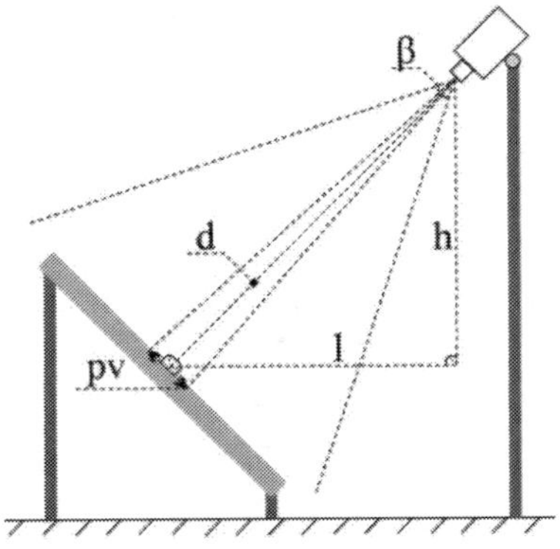

Fig. 2. Vertical field of view of the thermal camera to the photovoltaic panels.

Table 1. Standard IR camera resolutions and viewing angles.

Camera parameter	Lower class	Medium class	Professional class
Resolution	160 x 120	320 x 240	640 x 480
Available viewing angles	57° x 44° 55° x 43° 31° x 23°	45° x 34° 30° x 23° 24° x 18°	42° x 30° 34° x 24° 24° x 18°

If the PV system consists of many short PV panel rows, lined up one behind the other, then the camera must capture all of them. In such rare cases, the vertical viewing angle is also critical and its field of view also needs to be estimated in the same manner.

4 Calculation of the pixel size and pixel number per PV cell for certain camera and determining the right resolution

The current IR cameras offered on the market are most likely equipped with IR sensors with one of the following standard resolutions presented in Table 1. The higher resolution cameras cost more and the price matters. The PV systems are expensive and with comparatively low investment return ratio, which means that at least the IR diagnostic system must be with low enough price, but precise enough. Which IR camera resolution is best for certain case depends mainly on the size of the monitored object, the sensor field of view (FOV) and on the installation distance. The FOV depends on the horizontal (α) and vertical (β) viewing angles of the camera.

Once the needed camera viewing angles are determined, the next step is the right resolution of the IR sensor to be found in order each solar cell to be pictured with at least 5 pixels in each direction. The next step is the horizontal and vertical pixel sizes to be calculated for one of the 3 standard resolutions and for the chosen FOV. The best way is the first calculation to be done for the middle resolution 320 x 240. This means that the horizontal number of pixels is 320 and the vertical one is 240. First the instantaneous field of view (IFOV) of one pixel must be calculated for both horizontal (IFOVh) and vertical (IFOVv) directions, based on the horizontal α and vertical β viewing angles of the IR sensor, and based on the camera resolution, which consists of certain horizontal and vertical number of pixels. These parameters are the one pixel viewing angles.

$$IFOVh = \frac{\alpha}{horizontal\ number\ of\ pixels} \tag{5}$$

$$IFOVv = \frac{\beta}{vertical\ number\ of\ pixels} \tag{6}$$

If the camera is looking at 90° in both directions to the monitored object, the PV panels in this case, then the horizontal and vertical viewing planes consist of 2 right angle triangles, Figure 1 and Figure 2. For both of them are valid the following equations (7-10), where "ph" and "pv" are the pixel sizes in horizontal and vertical directions, and "dsc" is the distance from the IR camera to the particular solar cell for which the calculations are done.

$$tg\left(\frac{IFOVh}{2}\right) = \frac{ph/2}{dsc} \tag{7}$$

$$tg\left(\frac{IFOVv}{2}\right) = \frac{pv/2}{dsc} \tag{8}$$

$$ph = 2dsc \times tg\left(IFOVh/2\right) \tag{9}$$

$$pv = 2dsc \times tg\left(IFOVv/2\right) \tag{10}$$

If a standard solar cell with certain horizontal (sh) and vertical (sv) dimensions must be monitored, then the actual capturing number of pixels in horizontal (np_h) and vertical (np_v) directions are:

$$np_h = \frac{sh}{ph} \tag{11}$$

$$np_v = \frac{sv}{pv} \tag{12}$$

The problem with these calculations are that the horizontal and vertical viewing angles are not perpendicular to the monitored object, unless only one PV panel is monitored, in which case we could accept that at least the horizontal viewing angle is. This leads to a certain inaccuracy of the calculations, which in most practical cases is irrelevant. No ready to use tables could be presented with standard PV panel sizes and appropriate camera resolutions because the IR camera parameters depend on the number of monitored panels, and usually they are more than one, on the number of PV panel rows, on the viewing angles and installation distance. Each case is different and requires its own project to be created.

If the calculated number of pixels per direction is not enough for precise diagnostics, then the numbers for the higher IR camera resolution, but for the same viewing angles, which is twice higher, are twice higher. If the number of pixels is not enough again, then a camera with pan and tilt system might be selected.

5 Discussion and results

An example thermal image of the new PV plant located at University of Ruse is presented on Figure 3. It consists of 3 PV panel rows of 12 panels each with sizes 1 m wide and 2 m high. The thermal image proves that the PV panels do not suffer from any factory defects. The defects appear as hot spots. The only visible hot spot is on the 4th panel from right on the first row and it is caused by the shade which the IR camera casts over the first row.

Fig. 3. IR image of a PV plant at University of Ruse.

The IR image is taken with Konika Minolta thermographic camera Mobotix Mx-M16TB-R079, with infrared sensor with resolution 336 x 252 pixels and horizontal/vertical image angles of 45°/32°. The camera looks against the middle of the rows. Its installation distance to the middle of the 1st, 2nd and 3rd rows are 14.99 m, 20.83 m and 26.87 m, respectively. The distances to either end of the 1st, 2nd and 3rd rows respectively are 16.15 m, 21.68 m and 27.53 m.

Table 2. Calculated capturing number of pixels for the best and worst-case scenario.

Place of PV panel in a row	Number of pixels in a direction	Number of pixels per solar cell		
		1st row	2nd row	3rd row
Middle	Vertical	4.82	3.47	2.69
Either end	Vertical	4.47	3.33	2.62
Middle	Horizontal	4.57	3.29	2.55
Either end	Horizontal	4.24	3.16	2.49

Table 2 presents the calculations of the capturing number of pixels for the closest and furthest PV cells from each row of the presented PV plant, based on the presented methodology. The camera has two sensors and plays a double role. It is used for continuous security surveillance as well as for rough IR diagnostic of the PV panels. Two to five capturing pixels are enough for only rough diagnostic process, which is the required purpose in this case. In case of occurring damage, it will be investigated in more detail with another modular IR camera.

If the requirements towards the camera change and more detailed continuous IR imaging is needed, then there is no need for these calculations to be done again. Twice higher IR sensor resolution with the same viewing angles will lead to twice as many pixels per solar cell. Usually, calculations are needed only for the worst-case scenario, which is for the solar cell with farthest distance from the IR camera. In the presented case, these are the cells at either end of the 3rd row. The right IR camera resolution for PV plant diagnostics depends on the requirements and parameters of the project, and on the budget. The lowest class IR camera that allows precise enough diagnostic process is the best solution.

The IR diagnostic and the needed pixel count per solar cell calculations are not conceptual, but increasingly used nowadays. There are many extensive works describing the appearance of the different PV panel defects on the thermal images and the possible optimization technics, and this is not the aim of this work. At the same time, no methodology for the needed capturing pixel count calculation is found, but it has been developed and presented.

The IR imaging is used for new PV plant check for warranty purposes. Each PV panel has its warranty period and if a defect is found it could be replaced. Most of the defects are not visible with naked eyes. If damage in an out of warranty PV panel is found the optimization depends on the case. If the PV plant is installed in an urban environment, the PV panel must be replaced because it creates risk of fire and degrades the generated energy for that particular string, which will worsen the PV plant efficiency and the return of the investment. If the PV infrastructure is located outside of urban areas and if there is not any risk an occurring fire to create serious damage, then the defective PV panels could be combined in different strings, depending on the level of degradation of their output power. In such a case, they will continue to generate useful energy without degrading, but improving the PV plant performance and hence the return of the investment. PV plants have a low investment return ratio compared to other conventional energy ones and their functional optimization is critical.

6 Conclusions

The work presents a methodology for PV plant IR monitoring system design, based on calculations of the capturing number of pixels per PV cell. The correct sizing allows detailed inspection of the PV plant needed for timely prevention of fire hazards and for PV system optimization, which can improve the investment return ratio. The fire hazard must be minimized because otherwise it could impede the widespread use of the greener PV systems in urban environments, which is the optimal case.

The presented methodology for IR camera selection is based on the resolution of the IR sensor, the installation distance and the size of the monitored object. Its main drawback is that it does not correct the calculations for different than perpendicular horizontal and vertical camera viewing angles, which is the most common case. More often the camera cannot be installed high enough for perpendicular vertical viewing angle because it creates undesirable shade over the monitored PV panels. At the same time, it could be with close to perpendicular horizontal viewing angle for one PV panel, but it will not be with such for the next one.

These calculations are sharp enough in many practical cases, but in certain situations more precise results are needed. The improvement of this methodology for different horizontal and vertical viewing angles and how much they influence the pixel sizes is aim of future work, because it requires an extensive additional analysis. The farther the angles are from 90° the wider the pixels. Which parameter influences most the pixel size is also important for the precise design process of an IR diagnostic system and will be investigated in more detail in future works.

References

1. C. Genç, A. Sakalli, I. Stoyanov, T. Iliev, G. Mihaylov, I. Beloev, *Development of Wind Energy and the Installed Wind Power Plants in Turkey*, E3S Web Conf., **207**, 02013, (2020)
2. G. Gospodinov, K. Shtereva, V. Zhelyazova, V. Mutkov, Y. Neikov, V. Dimov, *Laboratory electronic circuit for tracing I–V and P-V characteristics of photovoltaic devices*, in Proceedings of the 10th International Symposium on Advanced Topics in Electrical Engineering (ATEE), Bucharest, Romania, pp. 57-61, (2017)
3. A. Boyadzhieva, B. Evstatiev, N. Mihailov, L. Mihailov, *Analysis of the Efficiency of a Stand-Alone PV System for Powering of a Detached House in the Village of Nikolovo, Bulgaria*, in Proceedings of the International Conference on Applied and Theoretical Electricity (ICATE), Craiova, Romania, pp. 1-4, (2021)
4. K. Simeonov, N. Mihailov, N. Valov and K. Gabrovska-Evstatieva, *Analysis of a PV Installation with a Battery Storage and BMS at a Residential Building*, in Proceedings of the 8th International Conference on Energy Efficiency and Agricultural Engineering (EE&AE), , Ruse, Bulgaria, pp. 1-5, (2022)
5. M. Isaied, A. A. Tarabsheh, *Modeling of Photovoltaic-Wind Hybrid Systems*, in Proceedings of the 2022 International Conference on Communications, Information, Electronic and Energy Systems (CIEES), Veliko Tarnovo, Bulgaria, pp. 1-6, (2022)
6. I. S. Stoyanov, T. B. Iliev, *An integrated approach for improvement of the efficiency of the photovoltaic system by using cogeneration*, Bulgarian Chemical Communications, **52**, C, pp. 46-52, (2020)
7. Z. Zarkov, V. Milenov, *DC-DC Converter for Adaptation of Thin-Film PV Panel I-V Characteristics for Microinverter*, in Proceedings of the 2022 International Conference on Communications, Information, Electronic and Energy Systems (CIEES), Veliko Tarnovo, Bulgaria, pp. 1-5, Veliko Tarnovo, Bulgaria, (2022)
8. L. V. Cárdenas, H. V. Flores, *Estimation of the Photovoltaic Potential on Rooftops in the City of San Pedro Sula, Cortés, Honduras*, E3S Web of Conf. **379** 03001 (2023)
9. S. Singh, N. Chander, Energy, **199**, pp. 351-367, (2022)
10. M. Oliveira, D. Cassini, A. Diniz, L. Soares, M. Viana, L.Kazmerski, V. Lins, Solar Energy, **191**, pp. 235-250, (2019)
11. I. Stoyanov, T. Iliev, G. Mihaylov, *Simulation Investigation of the Wind Load of Photovoltaic Panels*, E3S Web Conf. **327** 02002 (2021)
12. T. Yordanov, N. Mihailov, B. I. Evstatiev, *A Conceptual Model for Remote and Virtual Investigation of Photovoltaic Modules*, in Proceedings of the 8th International Conference on Energy Efficiency and Agricultural Engineering (EE&AE), , Ruse, Bulgaria, pp. 1-4 (2022)

13. I. Beloev, K. Gabrovska-Evstatieva, B. Evstatiev. Acta Technologica Agriculturae. **20**(4), pp. 85-90. (2017) https://doi.org/10.1515/ata-2017-0017
14. S. Letskovska, K. Seimenliiski, Journal of Computer Sciences and Communications, **9**, (2020). (in bulgarian)
15. G. Álvarez-Tey, R. Jiménez-Castañeda, J. Carpio, Infrared Physics & Technology, **87**, pp. 40-46, (2017)
16. Z. Yahya, S. Imane, H. Hicham, A. Ghassane, E. Safia, Sustainable Energy Technologies and Assessments, **52**, part A, p. 102071, (2022)

A model of a cyber-physical installation for smart greenhouse agriculture

Miglena Temelkova* and *Nikola* Bakalov

University of Telecommunications and Posts, Faculty of Telecommunications and management, 1 Acad. Stefan Mladenov Str., 1000 Sofia, Bulgaria

Abstract. The article reviews the main aspects, parameters and effects integrated by smart agriculture. An analysis of the main definitions of the "smart agriculture" concept is drawn, and, a cyber-physical installation for smart greenhouse agriculture is modeled on that basis. The specific benefits from the cyber-physical systems in the agricultural production process are also defined by taking into account the actual prerequisites, factors and indicators related to smart-based agriculture. The main restrictive conditions are generated, as well as the managing parameters and the criterion for optimality of the model of a cyber-physical installation for smart greenhouse agriculture. At the same time, the architecture of a cyber-physical installation for smart greenhouse agriculture is described, while there are the specific mathematical variables integrated in it from the optimization task set in the model. Highlighted are the specific activities performed by each structural element of a cyber-physical installation for smart greenhouse agriculture, where their practical-and-applied benefits and efficiency are also defined, which ultimately leads to achieving the set optimization criterion – maximum return of the invested financial resources.

1 Introduction

The relevance of the Concept for smart agriculture nowadays is related primarily to a basis of cumulatively available fundamental global needs – supplying the growing world population with foodstuff, efficient use and preservation of the global natural assets, climatic and nature-sustainable development of agriculture, meeting the requirements of Industry 4.0 for digitalization and smart-based development of agriculture as a sector of economy. A number of authors have researched the meaning of the "smart agriculture" concept as a starting point for reforming and restructuring the agricultural industry.

According to some authors [1], innovative technologies, such as artificial intelligence, the Internet of things and mobile internet provide realistic solutions for a number of challenges in the agricultural sector. Many authors [2, 3, 4, 5, 6, 7, 8, 9] assume that smart agriculture is based on the integration of information and communication technologies into machines, equipment and sensors related to agricultural production. On this basis, the combined application of information-and-communication technological solutions, such as

* Corresponding author: m.temelkova@utp.bg

accurate equipment, the Internet of things, blockchain, cloud technologies, sensors and driving mechanisms, geopositioning systems, large bulks of data, unmanned aircrafts and robotics are a prerequisite for achieving more efficient, more productive and more-sustainable agriculture. This accurate-based and resource efficient approach is influenced by the parameters of Industry 4.0 and enables the implementation of the cyber-physical systems in agriculture as a prerequisite for smart-based planning, management, resource return and yield efficiency.

In some publications [3], smart-based agriculture is determined as the latest agricultural revolution, which, through the Internet of things, connects smart machiness and sensors in farms, and makes the sustainable agricultural practices data-based. A number of publications [10] assume that a smart farm should be based on a blockchain security framework, which efficiently checks the status of the device and the sensor disorders, and mitigates threats. The researched authors also point out that a key point in smart agriculture is "optimization", which is aimed at optimizing each variable and input during the production stage [3]. The sustainability of agriculture using information technologies is reduced to increasing the quantity and quality of the manufactured production, and, at the same time – keeping the environmental balance, preserving the natural resources and avoiding their irreversible depletion.

According to some other authors [11], the digitalization of agriculture facilitates monitoring, marketing, technologies and accelerates the production process through smart and critical application of digital models and technologies.

Smart agriculture is a management concept focused on providing to the agricultural industry an infrastructure for using advanced technologies – including big data, cloud computing and the Internet of things – for following up, monitoring, automation and analyzing operations. [4]

The technologies, which are most often mentioned in literature [3, 4], and are applicable in smart agriculture, are the following:

- ➢ sensors for scanning the soil and managing irrigation, light, humidity and temperature;
- ➢ telecommunication technologies, such as GPS and GIS;
- ➢ hardware and software for specialized applications;
- ➢ digital tools for taking online decisions;
- ➢ smart forecasting;
- ➢ smart planning;
- ➢ smart management;
- ➢ robotics;
- ➢ automation.

It can be summarized on the basis of the carried out theoretical research of "smart agriculture" that it resolves the main problems related to sustainable development and environment preservation through optimal, but also environment-friendly and nature-sparing use of the soil, water and energy [12, 13]. In that sense, the main task in implementing cyber-physical systems in greenhouse production is observing the parameters of the core resources involved in manufacturing agricultural production:

- ➢ management of maintaining a healthy soil;
- ➢ irrigation management;
- ➢ pesticide use management;
- ➢ energy source management;
- ➢ management of the provided light;
- ➢ management of biodiversity in the soil and air;
- ➢ production process management.

The researches by a number of theoretical sources show that the sustainability of smart agriculture is a complex, multilayer concept, which sets as a goal achieving in parallel profitability and preservation of the global natural assets and ecological balance for the future generations.

2 Formulation of the main parameters of the Model of a cyber-physical installation for smart greenhouse agriculture

2.1 Parameters ensuring the optimality of the model

The integration of the cyber-physical system within the Concept for smart greenhouse agriculture requires the formulation of the main target parameters. Undoubtedly, the introduction of the philosophy of the cyber-physical systems in greenhouse production is aimed at achieving higher added value based on a more efficient decision making (in respect to expedience, resource consumption and profitability) and more efficient production operations (in respect to resource consumption and effectiveness), related to the exploitation of resources, data gathering and analysis in the course of the production process, the cultivation of the farm lands and harvesting the crop. That means to largely reduce the conventional manufacturing and labour processes on account of the automated, smart-based and smart-managed production-and-technological operations.

The restrictive conditions in the model of a cyber-physical installation for smart greenhouse agriculture result from the need to reduce resource consumption in the agricultural production process. The main production resources in agriculture are time (as duration of the entire production process), water (for irrigation), chemicals ensuring healthy and high quality production (fertilizers, spraying chemicals, etc.), energy (used for heating and air-conditioning), labour (in terms of costs for human resource remunerations), depreciation of the used technical equipment (in cash equivalent), penalty fines related to breaching the ecological and biological equilibrium (if that is the case). The aforementioned resources have their value expression, and, in general, determine the cost price for manufacturing 1 kg of agricultural greenhouse production.

The managing parameters in the model of a cyber-physical installation for smart greenhouse agriculture should be related to preliminary set borderline values in respect to the qualitative and quantitative characteristics of the soil, humidity, heat, illumination, biodiversity, as well as in respect to the borderline qualitative parameters for the content of pesticides in the agricultural production.

On the background of the defined restrictive conditions and managing parameters in the model of a cyber-physical installation for smart greenhouse agriculture, there could also be determined the optimality criterion – maximum return of the invested financial resources.

2.2 Structural parameters of the model

The determined target parameters of the model of a cyber-physical installation for smart greenhouse agriculture have their substantial contribution also for structuring the cyber-physical installation itself. It has a six-degree hierarchical structure, where each structural element is interrelated and determinating in respect to all the others. On this basis, the cyber-physical installation for smart greenhouse agriculture has the following structure (Figure 1):

- a greenhouse agricultural system, comprised of its two main subsystems – production and management one, where the physical manufacturing process takes place in a particular temporal and spatial aspect;

https://doi.org/10.1051/e3sconf/202340402004

- a sensor-smart system designed for gathering data from the real-life physical farm land through sensors, controlling apparatuses, technologies, management systems, which through the Internet of things maintain connection both with the real-life physical environment, and with the virtual one;
- an information system, where the signals gathered by the sensor-smart system are transformed into semantically rationalized information;
- a cyber system functioning as a central information centre, comprised of a network of cyber devices, which store a large bulk of information for different time periods, and with different situational variables, which analyze algorithms, models and instruments via specific methodologies that have been programmed in them, and generate, on this basis, alternative options for a choice of management decisions;
- a smart system generating in-depth knowledge about the physical greenhouse agricultural system based on expert systems, whereby a certain decision is analyzed and informationally supported;
- a management and feedback system, which implements the decision chosen by the system and configures the feedback from the cyber space to the physical greenhouse agricultural system.

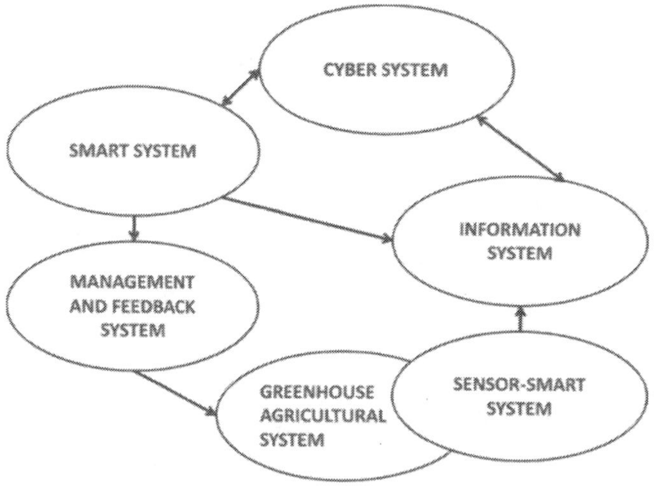

Fig. 1. Architectural structure of a cyber-physical installation for smart greenhouse agriculture.

At the same time, the described smart greenhouse agriculture is based on information and communication technologies. Thus, the model of a cyber-physical installation for smart greenhouse agriculture should include in its structure three interrelated and interdependent technological areas:

- management information systems;
- smart-based agriculture;
- process automation;
- robotized performance of certain activities.

The management information systems are associated with data gathering, processing, storing, analyzing and using by its designation on the basis of sensors, transducers and other technological solutions. It is these systems that allow adequate smart-based forecasting, planning, coordinating, controlling and managing the agricultural production processes. The management information systems draw their physical data from the soil, the air, from plants' root and foliar systems. This data is gathered by the sensor-smart system of the cyber-physical installation and is processed and transformed into digital information (by the information system of the cyber-physical installation), which is compared to the determined borderline values in the restrictive conditions defined in the model (by the cyber system of the cyber-physical installation). If necessary (when the measured physical data exceeds the indicated limits), in a smart-based way (through the management and feedback system of the cyber-physical installation) specific production-and-technological processes are activated, such as irrigation, heating or cooling, or ventilating, opening and closing the roof of the greenhouse, nourishing the soil with certain substances, treating the foliage against pests, etc.

Smart agriculture concerns managing the spatial and temporal variability to ensure the return of the invested financial resources with coordinated observation of no impact, or at least, reducing the impact on the environment. Critically important are the systems facilitating the decision making in view of ensuring optimality in respect to the return of the invested resources.

The automation of the agricultural production processes issues from the commands given by the smart system to the main robotized and automated elements of the cyber-physical installation.

Robotics (unmanned aircrafts for treatment, drones, etc.) ensures maintaining the parameters, which are set via the restrictive conditions of the model, and, on this basis, achieving the optimality criterion set as a goal.

3 A model of a cyber-physical installation for smart greenhouse agriculture

The model of a cyber-physical installation for smart greenhouse agriculture is related to a particular process organization, which is based on the parameters ensuring the optimality of the model and of its structural parameters. (Figure 2).

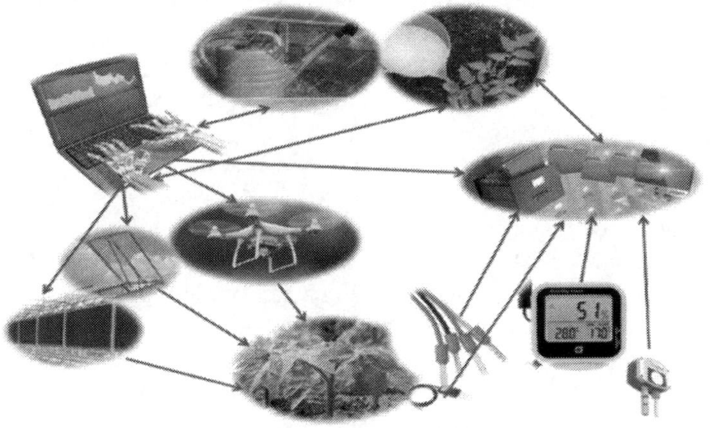

Fig. 2. A model of a cyber-physical installation for smart greenhouse agriculture. Author's own elaboration.

The sensor-smart system of a cyber-physical installation for smart greenhouse agriculture gathers via sensors and transducers data from the real-life physical greenhouse agricultural system about:

- ➢ the soil humidity;
- ➢ the minerals in the soil;
- ➢ the temperature of the air;
- ➢ the air humidity;
- ➢ the air circulation;

- plants' illumination;
- the condition of plants' foliage in view of the presence of a certain type of pests;
- the condition of plants' root system foliage in view of the presence of a certain type of pests.

The data gathered via the sensors and transducers is transmitted into the information system, where it is turned into semantically meaningful and relevant information. The synthesized information is transmitted from the information system into the cyber one, where a comparison is made with a big database, as well as an analysis of the condition of the physical greenhouse agricultural system. On the basis of the integrated restrictive conditions of the model in respect to the resources, the cyber system provides recommendations for the necessity of carrying out certain operational activities, such as fertilizing, irrigating, spraying against pests, heating or cooling the air, ventilating and other agricultural activities related to the normal development of the cultivated crops.

The smart system of the cyber-physical installation for smart greenhouse agriculture is based on the alternative options for a decision generated by the cyber system, where the latter provides approval for carrying out certain operational activities. At the same time, on the basis of the integrated therein managing parameters of the model of a cyber-physical installation for smart greenhouse agriculture, it defines accurately and precisely:

- the quantities of water necessary for irrigation and the duration of the irrigation process;
- the quantities of fertilizers or chemicals, whereby soil is enriched (if such a need is defined by the cyber system);
- the quantity of energy necessary for reaching particular temperature of the air (heating or cooling);
- the quantity of water, which has to be sprayed into the air in the greenhouse, or the time necessary for the greenhouse roof to remain opened (on the basis of information from a sensor in the indoor greenhouse environment about humidity outside of the greenhouse) until reaching the required air humidity;
- the extent of the opening of the roof or of other parts of the greenhouse in view of ensuring the necessary circulation of the air and illumination, providing for the respective period of the crops growth;
- the type and quantity of the substance necessary for treating plants' foliage or root system in the greenhouse.

Upon diagnosing the needs of the real-life physical agricultural environment, the smart system sends data about those needs also to the information system, which generates a new move to the cyber system, in order to check the result, which would have been generated when carrying out the activities determined by the smart system of the cyber-physical installation for smart greenhouse agriculture. Thus, the cycle of the cyber-physical installation is turned again, but solely and only in respect to the relation "information system – cyber system – smart system".

The management and feedback system, after completing the verification of the efficiency of the activities envisaged to be performed (the verification is on the basis of going through the cycle "information system – cyber system – smart system", ordered by the smart system), commands the automated and robotized systems and technologies to perform certain operational activities, such as:

- programming and starting the drip irrigation systems;
- programming and starting the robotized fertilizing equipment;
- programming and starting the air-conditioning units;
- programming and starting the installations for air humidification or opening to a certain extent the roof and/or other parts of the greenhouse in view of ensuring the

necessary humidity of the air on the basis of information from the external environment of the greenhouse;

- programming and starting the air circulation installations or opening to a certain extent the roof and/or other parts of the greenhouse in view of ensuring the necessary air flow on the basis of information from the external environment of the greenhouse;
- programming of the opening to a certain extent of the roof and/or other parts of the greenhouse and starting this process in view of ensuring the necessary and illumination of the crops;
- programming and starting drones for spraying the plants and/or their root system in view of protecting them from certain pests.

The management and feedback system of a cyber-physical installation for smart greenhouse agriculture follows the main parameters of the physical greenhouse agricultural system through the sensor-information system of the installation. If necessary, it will send commands to the latter to carry out specific activities. Thus, the entire process cycle of the cyber-physical installation for smart greenhouse agriculture is activated again.

4 Conclusion

Modelling a cyber-physical installation for smart greenhouse agriculture in an abstract and synthesized way summarizes the main parameters of a real-life example for a smart-based agricultural system. The precise definition of the restrictive conditions, the managing parameters and the criterion for optimality of the agricultural production, as well as the integration of those mathematical variables into the real-life architecture of a cyber-physical system adapted for the needs of the smart-based greenhouse agriculture, leads to achieving several major effects:

- reduced duration of the communication, management, planning and forecasting processes related to the functioning of the greenhouse;
- reduced time for carrying out particular agricultural activities;
- reduced loss of resources when performing certain agricultural activities, such as fertilizing, spraying, irrigating, heating, cooling, etc.;
- efficient growing of agricultural crops is ensured on the basis of accurate, precise and correct adherence to the technological processes and norms in agriculture.

The model of a cyber-physical installation for smart greenhouse agriculture leads to the following synergic effects, which provide grounds for the model to be upgraded and further improved:

- a rational production organization enabling the optimal capacity loading of both facilities and staff;
- optimal production costs ensuring the cost price per kilogram or generated yield;
- efficient production, technological and social added value ensuring the ecological and biological equilibrium, the preservation of the natural resources and the sustainability of production in a long-term period of time.

The aforementioned effects irrevocably determine also the achievement of the criterion for optimality set in the model of a cyber-physical installation for smart greenhouse agriculture – maximum return of the incurred financial expenses.

References

1. S. Mohamed, A. Belal, S. K. Abd-Elmabod, M. A. El-Shirbeny, Egyptian Journal of Remote Sensing and Space Science, **24**(4), (2021)

2. D. Pivoto, P. D. Waquil, E. Talamini, C. Spanhol, Information Processing in Agriculture **5**(1), (2017)
3. C. Bernstein, smart farming. Available at: https://www.techtarget.com/iotagenda/definition/smart-farming
4. M. J. O'Grady, G. M. P. O'Hare, Information Processing in Agriculture, **4**(3), pp. 179-187 (2017)
5. G. Idoje, T. Dagiuklas, M. Iqbal, Computers & Electrical Engineering, **92**, (2021)
6. A. A. Abubakar, J. Liu, E. Gilliard, Scientific and Practical Cyber Security Journal, **7**(1), pp. 38-53 (2023)
7. K. Hasan, Masriadi, Muchlis, A. Husna, Digital Farming and Smart Farming from the Perspective of Agricultural Students at Malikussaleh University 2022. In Proceedings of Malikussaleh International Conference on Multidisciplinary Studies (MICoMS), 3 (2022) DOI: https://doi.org/10.29103/micoms.v3i.230
8. A. L. Virk, M. A. Noor, S. Fiaz, S. Hussain, Smart Farming: An Overview. In: Patnaik, S., Sen, S., Mahmoud, M. (eds) Smart Village Technology. Modeling and Optimization in Science and Technologies, vol 17. Springer, Cham. https://doi.org/10.1007/978-3-030-37794-6_10
9. A. Rehman, Z. A. Shaikh, Smart Agriculture. In: Application of Modern High Performance Networks. Chapter: 6, pp.120-129 (Bentham Science Publishers Ltd. 2009)
10. S. Milovanovic, Ekonomika Poljoprivrede, **61** (2), pp. 471 (2014).
11. B. Güven, I. Baz, B. Kocaoğlu, E. Toprak, Smart Farming Technologies for Sustainable Agriculture: From Food to Energy. In: A Sustainable Green Future. pp. 481-506 (2023)
12. D. Tilman, K.G. Cassman, P.A. Matson, R. Naylor, S. Polasky, Nature, **418**, 6898, pp. 671-677 (2002).
13. L.A. González, G. Bishop-Hurley, D. Henry, E. Charmley, Anim Prod Sci, **54** (10), pp. 1687-1693 (2014)

 https://doi.org/10.1051/e3sconf/202340402005

Assessments regarding the performances of wind turbines from the roofs of buildings

Sanda Budea*

Department of Hydraulics, Hydraulic Machines and Environmental Engineering, University Politehnica of Bucharest, Romania

Abstract. In the current energy context, wind energy remains the main renewable energy source. Greater attention should be paid to small-sized wind turbines, mounted on the roofs of buildings, for their many advantages: easy to construct, do not require additional space, easy maintenance, eliminate losses from the system and transport costs, can provide alone or in addition to energy solar, the entire energy need of buildings, etc. The present paper analyzes three models of wind turbines that can be placed on the roofs of buildings: crossflow wind turbine Banki, ridge blade turbine and aeroMINE model. The analysis refers to their constructive solutions, the working principle, the optimal design for a good power coefficient and the energy that can be obtained with these small turbines for urban applications and not only.

1 Introduction

In the current energy context, there is a need for the accelerated development of energy production from renewable sources. Wind power remains the leading non-hydro renewable source, generating 1870 TWh in 2021, almost as much as all others combined. The amount of electricity generated by wind increased by almost 273 TWh in 2021, a 55% increase over that achieved in 2020 and the largest among all power generation technologies. Wind power generation increase by a record amount in 2021, but even faster growth is needed to get on the Net Zero Scenario trajectory. Wind power generation in the Net Zero Scenario, for 2030, could reach 8000 TWh [1].

An estimation of the increase in onshore wind power capacity by region can be found in Figure 1. In July 2021 the European Commission proposed to increase the bloc's renewable energy target for 2030 from 32% to 40%. The proposed target was further increased to 45% in May 2022. Many European countries have already expanded their renewables support mechanisms to accelerate capacity growth with a view to 2030 targets and in response to the energy crisis caused by Russia's invasion of Ukraine [1].

Recently, more attention has been paid to small-sized wind turbines, mounted on the roofs of residential or office buildings, having many advantages: they are small, light in construction and stand-alone, energy transport costs are eliminated, and the losses generated by the system. They can be mounted both horizontally and vertically, they do not

* Corresponding author: sanda.budea@upb.ro

require orientation in the direction of the wind, but the dominant direction of the wind is chosen. They have applications in urban areas by integrating them into the architecture of buildings or by retrofitting existing buildings. Low noise, good stability, and a cut-in velocity lower than 2 m/s, low maintenance costs, small area allocated compared to photovoltaic solar panels, are another advantages. There are also disadvantages: power coefficient below 0.3, the angle of the blades cannot be adjusted, inappropriate for multi-directional winds.

The present paper analyses three models of wind turbines that can be placed on the roofs of buildings: crossflow wind turbine Banki, ridge blade turbine and aeroMINE model. The horizontal mounting of these turbines is preferable for better stability and easier construction. The analysis refers to the constructive solutions, their working principle, the optimal design for a good power coefficient and especially the energy that can be obtained with these small turbines.

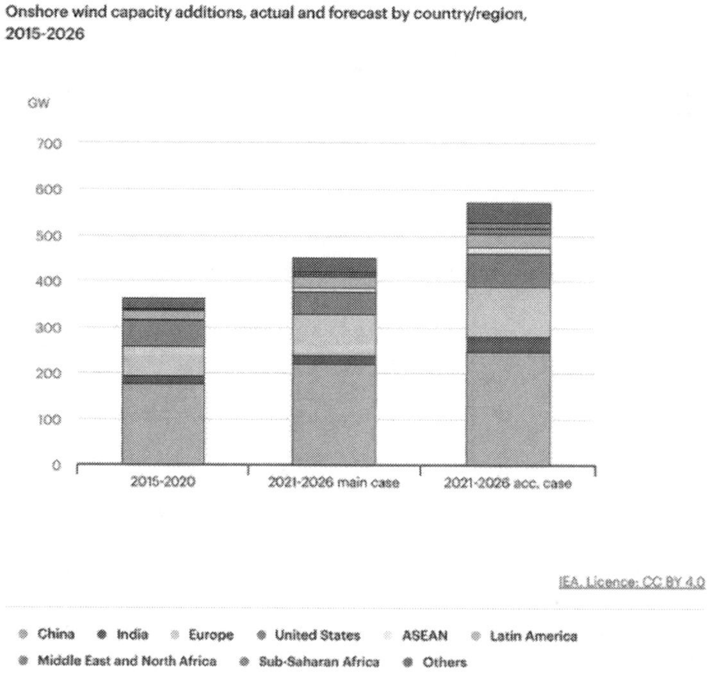

Fig. 1. Forecast by region regarding onshore wind turbine capacity additions [1].

2 Crossflow wind turbine

The multi-blade crossflow turbine has gained attention due to its high torque and low cut-in speed characteristics. This turbine is usually called the Banki wind turbine due to its similarity to the Banki water turbine [2]. The wind passes two times through the rotor blades, ensuring a better twisting moment. Moreover, the large number of blades gives the turbine the ability to have a low cut-in speed and a high operating torque.

The components of absolute wind speeds v, transport u and relative w for the first passage of the wind through the rotor are indicated in Figure 2. A good design assumes that the relative velocity w_2 passes through the center of the rotor, like in Figure 2, optimal angles are α_1=22°, β_1 between (38-45°), and β_2 = 90° [2]. Regarding the number of blades usually there are 6 to 24 blades, cylindrical shapes in single or multi-stages model. This number of blades depends on the rotor diameter ratio, and influence the efficiency, cut-in, and cut-off of the turbine.

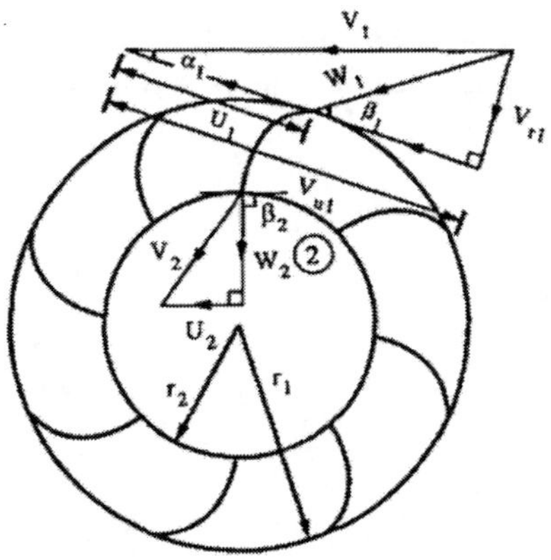

Fig. 2. Velocities triangles in section through the crossflow wind turbine.

Regarding the number of blades, numerical analysis and multiple experiments were done, and the conclusion was according to some authors [3] that 12 blades ensure a good power coefficient. According to other authors [4] [5], 16 blades with an angular extension of 60°, or 22 blades ensure good performances. Crossflow turbine for water has many blades 30-35 [6] for a good efficiency. An important parameter in the analysis of wind turbines is tip speed ratio *TSR*, like in relation (1), with *R* inlet radius, ω angular velocity, v wind velocity, n rotational speed:

$$TSR = \frac{R\omega}{v}; \tag{1}$$

$$\omega = \frac{\pi n}{30}. \tag{2}$$

The power extracted from the wind is:

$$P_w = \frac{\rho_{air}}{2} \cdot A \cdot v^3, \tag{3}$$

where A is the rotor area with diameter D and length L, as relation (4):

$$A = D \cdot L. \tag{4}$$

The turbine power is

$$P_t = c_p \cdot \frac{\rho_{air}}{2} \cdot A \cdot v^3. \tag{5}$$

So, the power coefficient depending on the wind direction (α) and *TSR* is:

$$c_p = \frac{P_t}{P_w}. \tag{6}$$

The torque coefficient is c_T and between the two coefficient that define the turbine performance is the *TSR*:

$$c_T = \frac{c_p}{TSR}. \tag{7}$$

Both the number of blades, as well as the profile of the blade and the characteristic angles are important. The TSR for this type of wind turbine is small, between 0.3-0.8. Good performances are obtained if guide vanes, casing and deflector are used, as in Figure 3 and Figure 4. These can lead to a c_{pmax}=0.265, or even 0.3. A complete analysis in this sense can be found in the paper [7].

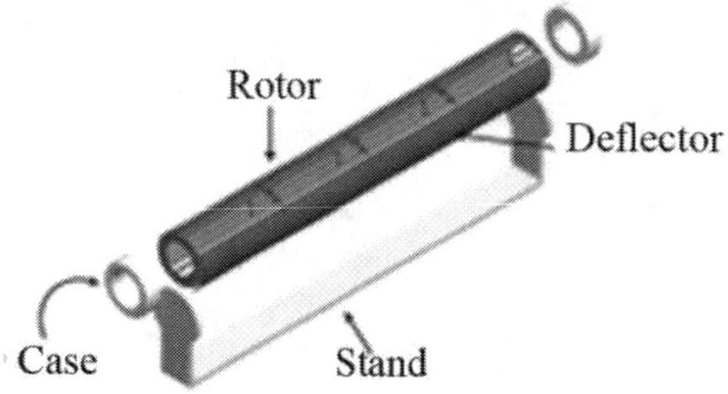

Fig. 3. Deflector for crossflow turbine.

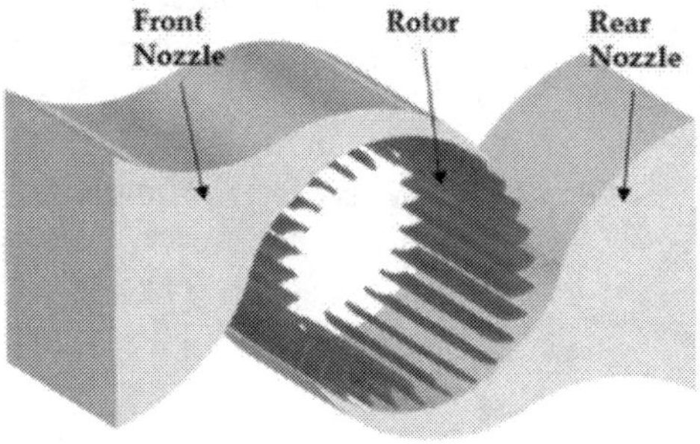

Fig. 4. Double nozzle [7].

New manufacturing technology involves 3D printing solutions for laboratory experimental models with 0.13 m diameter and 0.16 m lenght, easy to make, as in Figure 5. But be careful, the scale factor can greatly influence the performance of the turbines. With this small turbine can obtain a power of 7.35W assuming a wind speed of 6 m/s and a roof inclined at 45°.

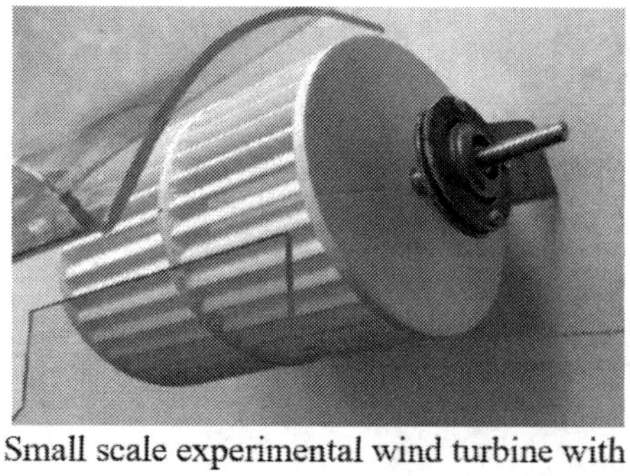

Fig. 5. Small scale experimental wind turbine with 0.13 m diameter and 0.16 m length.

3 Wind turbines on the ridge of the roof

A new solution, patented in 2010 by Raymond E. Paggi [10], refers to the mounting of small wind turbines, with a horizontal axis, on the ridge of sloping roofs. The rotor design is similar to a paddle wheel, with many blades on a horizontal shaft and a housing that includes air guide elements. Through the impact of the wind with the sloping surface of the roof, an important increase in the speed of the air reaching the turbines is obtained. The effect is like an airfoil, low pressure favors high speed according to Bernoulli's relation.

The horizontal mounting of these wind turbines on roofs is relatively easy. Figure 6 represents the effect of the sloping roof on the wind speed and in Figure 7 is presented a residential application. The system can be retrofitted to an existing sloped roof, mounted directly to the roof, or it can be included into the roof of a new building to improve wind energy capture.

The wind speed at the ridge of the roof increases significantly. The American specialists have shown that the optimal angle of inclination of the roof for a good multiplication of the speed, respectively the energy is 45° (Table 1). At a roof inclination of 30 °, the speed multiplication is 1.5 times [9].

Table 1. Wind velocity multiplier [9].

Roof pitch angle	0°	8°	15°	30°	45°	60°
Multiple of wind velocity	X 1.0	X 1.1	X 1.2	X 1.5	X 2.2	X 1.3

In terms of performance, turbines mounted on the ridge of sloping roofs can provide 5 to 8 times more energy compared to similar turbines mounted on a flat surface. According to [10], with a rotor with a diameter of 1.5 m and a length of 6 m (from 5 rotors x 1.2 m), swept area is 1.8 m2, a power of 100 W is obtained at 4.5 m/s wind speed for a flat location. Due to the mounting on the slope, there is an important factor of multiplying the wind speed by 1.5-2.2 times (6.75-10 m/s), resulting in powers up to 1500 W. See Table 2 adapted from paper [10].

Table 2. The effect of the ridge montage of the wind turbine adapted from [10].

Diameter 1.5 m	Wind velocity (m/s)	Area (m2)	Roof effect factor	Wind energy available (W)
Roof ridge WT	4.5	1.8	1.0	100
Roof ridge WT	4.5	1.8	1.5	337
Roof ridge WT	4.5	1.8	2.0	800
Roof ridge WT	4.5	1.8	2.2	1065

Fig. 6. The effect of the sloping roof on the wind speed [9].

Fig. 7. Ridge blade applicatins on house [8].

Small wind turbines, with an horizontal axis, are often located on the roofs of houses in the UK, Figure 7 or in industrial applications, like in Figure 8 [9].

Fig. 8. The ridge blade wind turbine industrial application [9].

If the rotor is placed on dominant wind direction, Ridge roof wind turbine has advantages likes: -works well even in turbulent regime, - works at optimal parameters from low wind speeds of 4.5 m/s - they are silent in operation - they are self-autoregulating - works well even in storm conditions. The blades are symmetrical, designed as a bi-directional turbine system and generate energy even if the wind changes direction.
When operating in conditions of a 40° inclined roof, wind speed of 6 m/s, with 5 rotors of 1.2 m length (6 m) and 1.5 m diameter, a power of 1 185 W results, i.e. a monthly energy of 425 kWh /month, more than 5 MW/year, if the wind blows half the time.

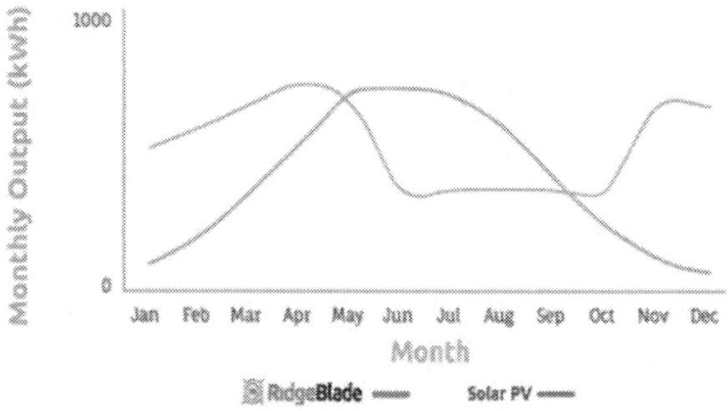

Fig. 9. Complementary of solar and wind energies [9].

Due to the fact that wind energy and solar energy are relatively complementary, the wind blows stronger at night, we have sun during the day, wind-solar hybrid projects have been implemented on the model of ridge blade turbine, like in Figure 9.

Such a project is carried out for a commercial building in Parc del Forum, Barcelona, with 20 ridge roof wind turbines, which at an average wind speed of 4.3 m/s, produce 14,000 kWh/year, and a solar PV installation of 20 kW, which at the average solar radiation of 185 W/m2 would annually produce 26,000 kWh/year [9].

4 Aeromine model bladeless wind turbines

A relatively recent model of wind turbines, small in size, without blades, mounted on the roofs of residential or office buildings, is the third comparative option that I propose in this study. The constructive solution of these wind turbines that can be placed on roofs, not in the ridge area like a ridge blade, but at the bottom edge of the roofs, is based on the principle of the Venturi channel: when fluids flow in a narrowed section, a baffle is formed between two profiles, increasing the speed of air flow. The depression created sucks the air up.

Fig. 10. The working principle of aeromines [11].

According to Berboulli's relation of conservation of energy, the decrease in pressure leads to the increase in speed, which favors the more accelerated movement of air upwards, through the two profiles, mounted face to face, like a Venturi channel. The turbine has double suction, the fluid flow in the internal circuit stimulates the external flow, between the 2 profiles in the form of a Venturi channel. The air flow passes through a turbine / propeller located at the bottom of the installation, generating energy [11], like in Figure 10. System named Aeromine begins operating around a wind velocity of 2.5 m/s. With wind velocity of 9-10 m/s and greater generated by Venturi channel effect, results an efficiency of 25%.
The equations are conformed [12]. The drop pressure from in to out of the foils is:

$$\Delta p = \frac{\rho}{2} \cdot c_{pres} \cdot U_{\infty}^{2}. \tag{8}$$

The mechanical / pneumatical power can be calculated with:

$$P = \Delta p \cdot A_d \cdot u_d \tag{9}$$

The wind power is:

$$P_w = \frac{\rho}{2} \cdot A \cdot U_{\infty}^{3}; \tag{10}$$

$$c_p = \frac{P}{P_w}. \tag{11}$$

The notations are: U_{∞} - wind velocity, c_{pres} – pressure coefficient, ρ – air density, A_d - area of the internal ducts, u_d – velocity in ducts, A – area of the rectangular section from the foils.

So, in experimental research from [13], the power coefficient for a wind turbine at 1/3 scale is 0.18 and from the real turbine with 3 m tall and 1 m profile chord, power coefficient is maxim 0.25 for wind velocity in channel 9 m/s [12]. The experimental tests were made with a foil profile S1210.

The operation of the system starts from wind speeds of 2.5 m/s, but in Venturi channel increases to 9 m/s . The internal propeller is not really a wind turbine, it is like a fan in reverse operating. Placed on the dominant direction of the wind, 1.5-2 kW can be obtained with a single wind turbine. The technical solution is relatively recent, other experimental studies follow to confirm the results. Aeromine is a bladeless wind energy solution which occupies only 10% of the area occupied by PV solar panels that provide similar power and can be integrated in residential or office buildings. Also in this case, solar PV-aeromine hybrid solutions are suitable.

5 Experimental results

A model of an aeromine turbine - with 2 profiles simulating a venturi channel for air I built in the laboratory. The profile is 0.5 m high and the chord of the profile 0.52 m, the inlet section 0.5 x 0.15 m, and in the narrowed area 0.5 x 0.05 m. With an anemometer with ±2% accuracy I measured the velocities of the air current in the reduced area and represented the values with the Sigmaplot 14 application, as in Figures 11a, b.

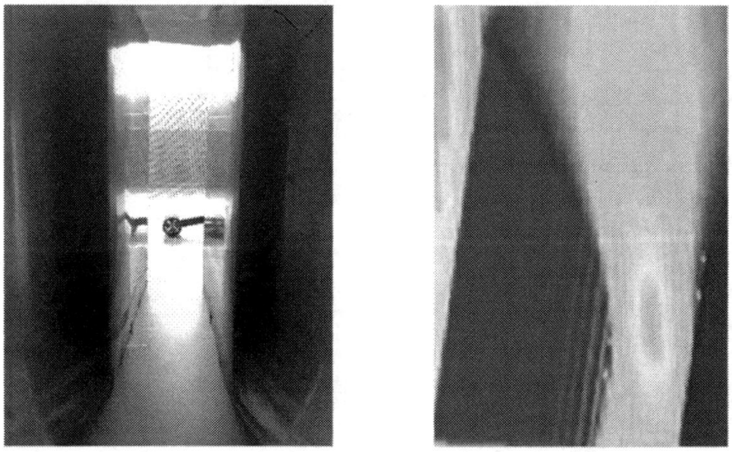

(a) (b)

Fig. 11. The experimental stand with aeromine profiles a) with anemometer, b) velocities spectrum.

Following the experimental measurements, it was observed that at relatively low velocities, below 4 m/s in the downstream area, in the narrowed area the velocity increases by a maximum of 3 times, but at speeds of 4.5-10 m/s downstream (Table 3), in the reduced area the speed increases less , 2.5-1.5 times, the current of air tends not to enter in the narrowed area, but to go around the outside of the profiles, see Figure 11b.

Table 3. Elements of comparison

Tip of WT	Wind velocity (m/s)	Cp max	Mounting	Area (m2)	Wind energy available (W)	Specificity
Crossflow	4.5 to 9	0.265	Horizontal	6x1.5	1200	2 passing
Roof ridge	4.5x2.2	NA	Horizontal	1.8	1065	Good power ratio W/ m2
Aeromine	2.5 to 9	0.25	Vertical	3	1500	Double suction

6 Conclusions

From the analysis of the three models of wind turbines that can be placed on the roofs of buildings: crossflow wind turbine Banki, ridge blade turbine and aeroMINE model we can draw the following conclusions:

i) They have many advantages - They are the most suitable for urban layouts, not affecting additional spaces. - Horizontal mounting of these turbines is preferable for better stability and easier construction. Exceptions are aeroMINE that are mounted vertically. - they are stand-alone, eliminating energy transport expenses and losses generated in the system. - does not require orientation to the direction of the wind but are mounted on the dominant direction of the wind. - They can be integrated into the architecture of new or existing residential or office buildings. - Low noise, good stability, does not create visual discomfort. - the cut-in velocity lower than 2 m/s, low maintenance costs, - they can replace or supplement the PV solar energy, from the roofs of the buildings, in this case they can fully ensure the energy needs of the buildings.

ii) Disadvantages - Relatively low power coefficient, below 0.3 - The blades are not adjustable - The resulting power is low, 2-3 kW per module.

iii) Additional observations:

There is a need for greater promotion of these turbine models for a cheap source of energy, right at the point of consumption - residential and office or industrial buildings.

- additional studies are required for the aeromine variant.
- It is necessary to adopt solar PV-wind hybrid solutions for more energy, but also for the fact that the presence of wind turbines in the vicinity of the PV panels ensures a better operation for both
- it is known that cooling panels have better performance, at the same time concentrating the heat drives air movement towards the turbines.
- The scale factor between the model and the real turbine must be taken into account, in the analysis of their performances.

The analysis should be completed with experimental results, in future detailed research.

References

1. https://www.iea.org/topics/net-zero-emissions (2023)
2. Al-Maaitah, A. Ayman, *The design of the Banki wind turbine and its testing in real wind conditions*, Renewable Eneryy, **3**, No. **6**, 7, pp. 781-786, (1993)
3. I. Ushiyama, N. Isshiki, G. Z. Chai, *Experimentally evaluating the design and performance of cross-flow wind rotors.* Energy and the Environment in the 1990s, 1st Worm Renewable Energy Congress (Edited by A. A. Sayigh). Pergamon Press, Oxford (1990)
4. A. Wibowo,; D.D.D.P Tjahjana,. B. Santoso, M.R.C. Situmorang, *Study of turbine and guide vanes integration to enhance the performance of cross flow vertical axis wind turbine.* AIP Conf. Proc. 2018, **1931**, 030043 (2018) [CrossRef]
5. T. Pujol, A. Massaguer, E. Massaguer, L. Montoro, M. Comamala, *Net Power Coefficient of Vertical and Horizontal Wind Turbines with Crossflow Runners.* Energies **11**, 110 (2018)
6. R. Adhikari, D. Wood, *The Design of High Efficiency Crossflow Hydro Turbines: A Review and Extension* , *Energies* , **11** (2), 267, 1-18 (2018)

7. Z. Sefidgar, A. Ahmadi Joneidi, A. Arabkoohsar, *A Comprehensive Review on Development and Applications of Cross-Flow Wind Turbines*. Sustainability, **15**, 4679, https://doi.org/10.3390/su15054679 (2023)
8. https://www.youtube.com/watch?v=eUv3Pt9LW7c (accessed April 2023)
9. The power collective, RB, https://www.switchontario.ca/resources/Documents/Managing%20Energy%202017/Managing%20Energy%202017%20-%20Business%20-%20RidgeBlade%20-%20Llion%20Rowlands.pdf Accessed April 2023
10. R. E. Paggi, *Roof ridge wind turbine*, United States Patent Application Publication Paggi, US 2010/0126086 A1, (accessed 27 May 2010)
11. https://www.aerominetechnologies.com/ (accessed in April 2023)
12. B. C. Houchens, D. V. Marian, S. Pol and C. H. Westergaard, *High Wind Speed Performance of AeroMINE at Pilot-Scale*, J. Phys.: Conf. Ser. **2265** 042065, IOP Publishing, (2022)
13. S. Pol, B. C. Houchens, D. V. Marian, C. H. Westergaard, *Performance of* AeroMINEs *for Distributed Wind Energy*, AIAA SciTech (2020)

Effects of exhaust gas recirculation (EGR) rates on emission characteristics of ethanol and methanol diesel blended fuels

Simeon Iliev*

University of Ruse, Engines and Vehicles Department, 8 Studentska Str., Bulgaria

Abstract. The introduction of ever more stringent diesel emission standards is associated with a reduction in greenhouse gases. These limitations push many researchers working in this field to look for alternative solutions with the aim of lower greenhouse gas emissions. One of the alternatives is the use of alternative fuels and additives in order to reduce the consumption of fossil fuels. The use of alternative fuels such as ethanol, methanol and others in combination with traditionally used technologies such as exhaust gas recirculation (EGR) is one of the solutions that would lead to the limitation of diesel engines emissions. The aim of the present study is to investigate the performance of a diesel engine with ethanol and methanol additives combined with different degrees of recirculation on engine emissions. For this purpose, a model of a diesel engine with EGR with the possibility of working with ethanol and methanol additives has been developed.

1 Introduction

Depletion of fossil fuel reserves, global warming and environmental protection worldwide encourage the search for alternatives to traditional fuels (diesel fuel and gasoline). As a consequence of the three global problems, the search and development of new technologies or the improvement and implementation of the old technologies known to us are stimulated in order to reduce the consumption of fuels and thereby the emissions released by internal combustion engines (ICE). Despite the strict measures that are imposed in connection with the protection of the environment, the demand for fossil fuels continues to grow provided that it is well known that alternative fuels can be an alternative to fossil fuels. Some sources report that more than 84% of CO2 emissions released into the atmosphere are from fossil fuels [1].

Despite the great restrictions imposed in recent years on the use of diesel engines, they are still widely used in various fields of industry such as shipbuilding, agriculture, transport (land, sea and rail), electric generators, construction equipment and tools. The reason for this wide application is due to their high reliability, fuel economy, adaptability and thermal efficiency. Due to the large number of vehicles and wide application, the share of greenhouse gas emissions emitted by diesel engines is also increasing. The pollution in big

* Corresponding author: spi@uni-ruse.bg

cities is especially great, causing acid rain, numerous respiratory and cardiovascular diseases and photochemical smog [2].

In order to reduce harmful emissions emitted by diesel engines, combinations of different technologies are used to control the combustion process in combination with the use of different alternative fuels. Exhaust gas recirculation (EGR) technology has been widely and successfully applied to both diesel and gasoline engines to reduce NO_X emissions. Many scientists are of the opinion that applying up to 15% of the EGR NO_X concentration can be reduced up to 80%. EGR can also be successfully used in diesel engines running on various alternative fuels such as biodiesel, various alcohol fuels, natural gas, hydrogen and others without deteriorating their thermal efficiency, and without a significant increase in fuel consumption. In one of the studies [3] con-ducted with a diesel engine and 10% EGR, a reduction in NOx and a slight increase in the PM and soot emission were found. In another study [4] conducted with a direct injection diesel engine using biofuel blends applied up to 20% of EGR was found to reduce NOx emissions and improve combustion characteristics.

1.1 Properties of Methanol and Ethanol

Methanol (methyl alcohol) and ethanol (ethyl alcohol) can be directly used without any engine modifications as a single-fuel and mixed with both diesel fuel and gasoline. They also have similar chemical properties (Table 1). Both alcohols have a simple chemical structure and are liquid and homogeneous [5]. They belong to the group of light alcohols [6].

Table 1. Comparison of fuel properties.

Properties	Euro Diesel	Methanol	Ethanol
Chemical formula	$C_{14}H_{28}$	CH_3OH	C_2H_5OH
Molecular weight (kg/kmol)	196	32	46
Flame speed rate (cm/s)	33	35	39
Boiling temperature (℃)	190–280	64.7	78.1
Density (kg/m3, at 20°C)	840	796	789
Flash point (℃)	52	11	13
Autoignition temperature (℃)	254	464	423
Lower heating value (MJ/kg)	42.74	20.27	28.40
Cetane number	56.5	4	6
Carbon content (wt.%)	86	37.5	52.2
Oxygen content (wt.%)	0	50	34.8
Heat of vaporization (MJ/kg)	0.27	1.11	0.92

Table 2 show diesel-methanol with 10% CH_3OH (M10) and diesel-ethanol 10% C_2H_5OH (E10) blends used for the analysis.

Table 2. Composition of diesel-methanol and diesel-ethanol blends.

Properties	**Diesel**	**M10**	**E10**
Diesel Fuel (%)	100	90	90
CH_3OH fuel (%)	0	10	10
C_2H_5OH fuel (%)	0	10	10
Density (kg/m^3)	840	836	835
Lower heating value (MJ/kg)	42.74	40.2	40.8
Heat of evaporation (KJ/kg)	270	357	334
Oxygen content (%)	0	5	3.1

C_2H_5OH and CH_3OH have a lower viscosity compared to diesel fuel (Table 1). This results in easier atomisation and evaporation and therefore better air-fuel mixing. These two alcohols do not contain sulphur unlike diesel fuel, their use as fuels in internal combustion engines contributes to the reduction of sulphur dioxide (SO_2) and to reduce acid rain, as SO and NOx emissions lead to acidic acid deposition. Due to the higher latent heat of evaporation of C_2H_5OH and CH_3OH compared to diesel fuel, it is associated with lower combustion temperatures, which in turn leads to lower NOx emissions. The content of oxygen in the molecules of C_2H_5OH and CH_3OH leads to a reduction in the formation of soot and particulate matter during combustion. This has a beneficial effect on the combustion process, which also improves. The combustion process of these two alcohols is associated with a high laminar flame propagation speed, which leads to an earlier completion of the combustion process and improves the thermal efficiency of the engine. High evaporative cooling leads to cooling of the combustion chamber during the intake and compression stroke and to an increase in engine volumetric efficiency, thus reducing the necessary work in the compression stroke. This cooling effect can also reduce NO_X concentration in the combustion chamber.

As some of the disadvantages of C_2H_5OH and CH_3OH are:

The cetane number of C_2H_5OH is low, and when mixtures between methane, C_2H_5OH and diesel fuel are used, it is necessary to add a cetane number improver to improve the properties of the mixture. In addition to the cetane number improver, an emulsifier is also added. This further increases the price of the mixture [7].

C_2H_5OH and CH_3OH have limited solubility in diesel fuel, which leads to phase separation of mixtures with these alcohols. The dynamic viscosity of C_2H_5OH is much lower than that of diesel fuel, which impairs the lubricating properties of the mixture. This is a problem because the lubricating properties of diesel fuel are used to lubricate the combustion system [8].

Both alcohols have low self-ignitability due to their high latent heat of vaporization, low cetane number, and high ignition temperature [9].

Thanks to the specific properties of C_2H_5OH and CH_3OH, they can be used in the form of mixtures in diesel engines without requiring structural changes to the fuel system, but the ratio of alcohols to diesel fuel must be low. In order for them to be used in their pure form, it is necessary to make structural changes to the fuel system.

Various studies have been done with mixtures of C_2H_5OH and CH_3OH. A study [10] involving a diesel engine and blends of biodiesel, C_2H_5OH, CH_3OH, and vegetable oil with diesel fuel, where each blend contained 15% alternative fuel by volume, discovered that the use of C_2H_5OH and CH_3OH blends led to decreased brake power, increased fuel consumption, and reduced CO emissions. In a separate study [11], an automotive common-rail diesel engine was utilized, fuelled by a blend of diesel and C_2H_5OH with a 20% C_2H_5OH content. The engine was operated under various working conditions, and the results demonstrated a reduction in soot emissions across all operating conditions. However, the thermodynamic efficiency was found to decrease with increasing C_2H_5OH content at low load conditions, while it showed a slight improvement at medium-high load conditions. Furthermore, in another investigation [12], the impact of different mixtures containing C_2H_5OH and diesel fuel (5%, 10%, and 15% C_2H_5OH) on the performance and emissions of a diesel generator was examined. The generator was subjected to a load range of 5 to 37.5 kW. It is characteristic of the experiment is that the fuel mixtures are directly injected into the combustion chamber of the engine. The obtained results show that in C_2H_5OH mixtures at low loads the in-cylinder peak pressure and heat release rate decreased and at high loads they increased. Increasing the C_2H_5OH content in the mixture results in increased ignition delay and reduced combustion duration. The temperature of the exhaust gases has also decreased. Increasing the C_2H_5OH content has also led to a reduction in CO_2 emissions by up to 8.6%. CO, HC and NOx emissions vary depending on engine load and C_2H_5OH content of the mixture. Other authors [13] studied diesel-methanol blends and solvents for better mixing of diesel fuel with CH_3OH. The performance and emissions at different fuel injection angles were investigated. The obtained results show that with the increase of the fuel advance angle, the thermal efficiency of the engine increases and the BSFC decreases. It was found that the NOx emissions increase with the increase of the advance angle of the fuel supply, with the increase of the CH_3OH content, the NOx emissions are decreasing. The effect of reducing NOx emissions is more noticeable at high loads. As the fuel advance angle increased, the CO content also decreased and the amount of smoke was reduced for the different blends.

2 Engine simulation

2.1 Simulation setup

The primary objective of this research was to develop a 1D model of a direct injection diesel engine. The model was designed to investigate the impact of fuel blends containing C_2H_5OH and CH_3OH with diesel fuel on both emitted emissions and engine specific fuel consumption. The specialized software tool AVL BOOST was utilized to construct the one-dimensional model. The model's accuracy was ensured through calibration, and the engine's configuration is depicted in Figure 1, while the specifications are presented in Table 3.

Table 3. Library with the used elements.

Engine Data	Value
Cylinders	4
Displacement	3666 cc
Engine Bore	100 mm
Engine Stroke	110 mm
Compression ratio	17:1

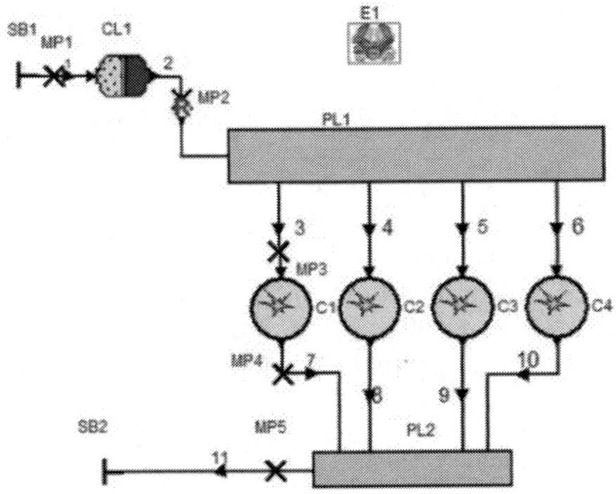

Fig. 1. Layout of a four-cylinder diesel engine.

3 Results

The current study focused on investigating the performance and emission properties of a four-cylinder, four-stroke diesel engine. The engine was operated using various mixtures of CH_3OH, C_2H_5OH, and pure diesel fuel, along with the implementation of exhaust gas recirculation (EGR). The AVL BOOST software was utilized to simulate and analyze the engine behaviour. The experiments were conducted at different engine loads and speeds. Specifically, the following fuel blends were utilized: 10% CH_3OH (or C_2H_5OH) and 90% pure diesel fuel, denoted as M10 (or E10) by volume.

3.1 Performance characteristics

Figure 2 illustrates the outcomes of brake specific fuel consumption (BSFC) achieved through a combination of alcohols and pure diesel fuel, along with exhaust gas recirculation (EGR).

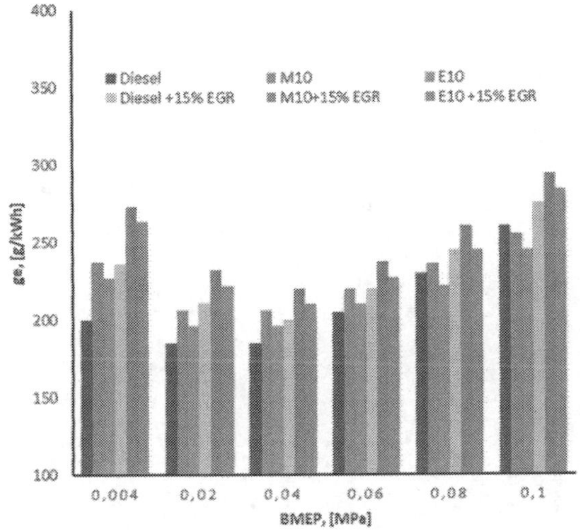

Fig. 2. Influence of E10, M10 and EGR on BSFC under various engine loads.

As can be seen from Figure 2, BSFC increases with increasing C_2H_5OH and CH_3OH content in the fuel mixtures. This can be explained by the lower energy content of CH_3OH (20.27 MJ/kg) and C_2H_5OH (28.40 MJ/kg) compared to diesel fuel (42.74 MJ/kg). In addition, the heat of vaporization of C_2H_5OH (0.92 MJ/kg) and CH_3OH (1.11 MJ/kg) is higher than that of diesel fuel (0.27 MJ/kg). As a consequence, C_2H_5OH and CH_3OH will absorb more heat in their vaporization process compared to diesel fuel, and this will lead to a drop in the temperature in the engine cylinder, which will be especially affected at low loads. This will lead to deterioration of the combustion process and in-crease in fuel consumption. A significant increase in fuel consumption was also observed for the fuel mixtures combined with a 10% recirculation rate. This can be explained by a deterioration of the combustion process due to a lack of oxygen due to the exhaust gases entering the cylinder from the EGR.

3.2 Engine emissions

The CO and HC emissions by blended alcohols and net diesel fuel and EGR are shown on Figure 3 and Figure 4.

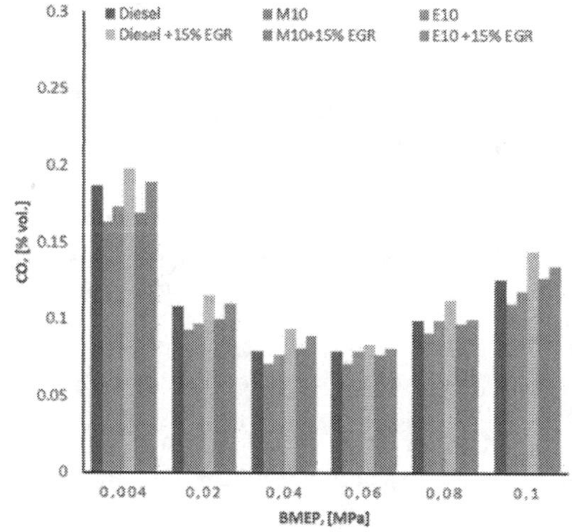

Fig. 3. Influence of E10, M10 and EGR on CO emissions under various engine loads.

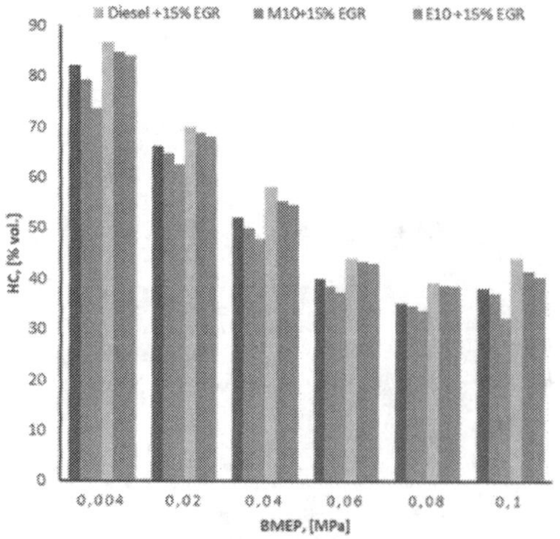

Fig. 4. Influence of E10, M10 and EGR on HC emissions under various engine loads.

The obtained results show that when using EGR technology, an increase in CO and HC emissions is observed for all used fuels and fuel mixtures. The CO content is higher in fuels that do not contain oxygen in their molecular structure. The reason for the obtained results is a deterioration of the combustion process due to dilution with inert gases. The higher oxygen content of alcohols contributes to the lower levels of CO emissions. Some of the reasons for the formation of HC emissions are the properties of the fuel, the presence of oxygen in the composition of the fuel, the combustion temperature, etc. The addition of C_2H_5OH and CH_3OH to diesel fuel results in a higher temperature in the cylinder, resulting in the fuel reacting more easily with oxygen. At the same time, CH_3OH and C_2H_5OH increase the laminar flame speed compared to diesel fuel, leading to a reduction in burn duration but an increase in combustion temperature and more complete combustion and lower HC emissions. The graphs show that this effect is most pronounced with C_2H_5OH - diesel blends. The flame speed of C_2H_5OH is 39 cm/s compared to CH_3OH 35 cm/s and diesel 33 cm/s.

The results for NOx emissions obtained by blended alcohols and net diesel fuel and EGR at different engine speed are shown on Figure 5, Figure 6 and Figure 7.

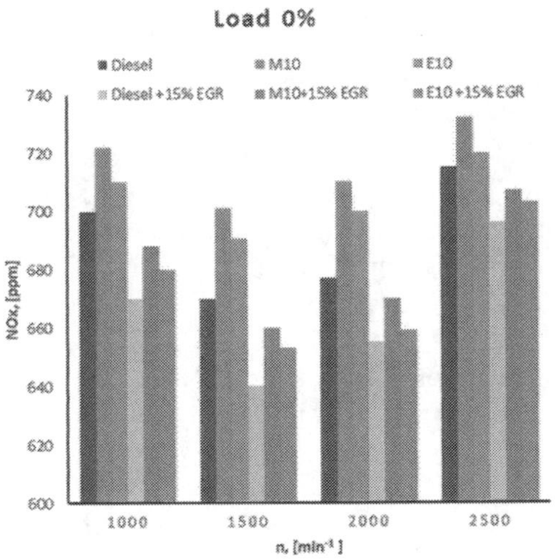

Fig. 5. Influence of E10, M10 and EGR on NOx under various engine speeds with no load.

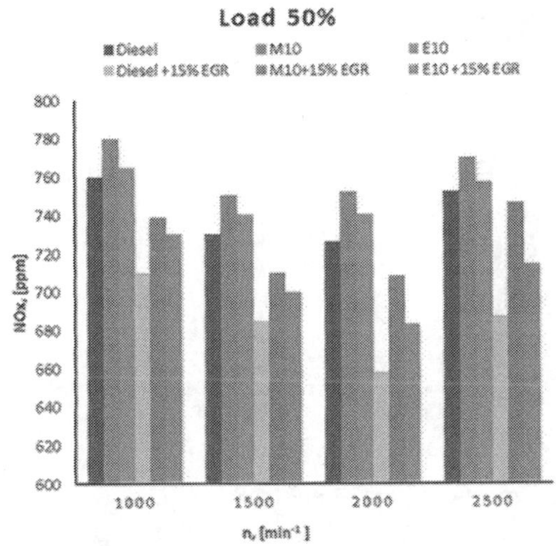

Fig. 6. Influence of E10, M10 and EGR on NOx under various engine speeds with 50% load.

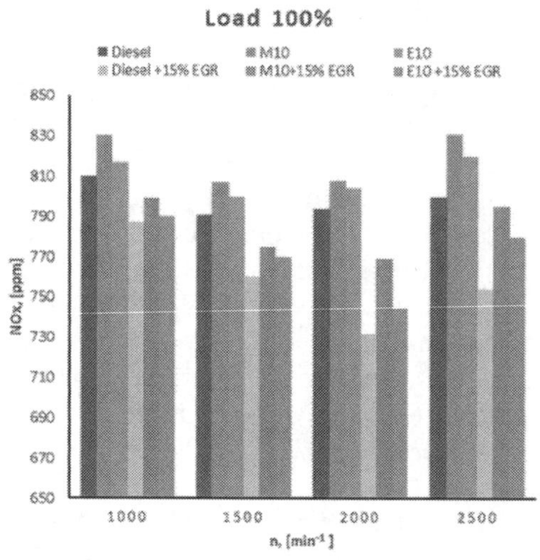

Fig. 7. Influence of E10, M10 and EGR on NOx under various engine speeds with 100% load.

The NOx emissions mainly depend on two factors: the presence of free oxygen and high in-cylinder temperature. The use of alcohols and alcohol additives can lead to two effects. The first is that the oxygen content of alcohols can enhance the formation of NOx. The second, due to their higher latent heating, reduces the combustion temperature in the combustion chamber, as a result of which the formation of NOx is reduced. Consequently, NOx increases with increasing alcohol content in the fuel mixture. It can be seen from the graphs that the increase in NOx is more significant in mixtures with CH_3OH. When the load increases, NOx emissions also increase. NOx reduction with EGR at high loads conditions are greater compared to no load.

The NOx emissions by blended alcohols, diesel fuel and EGR at injection timing change are shown on Figure 8.

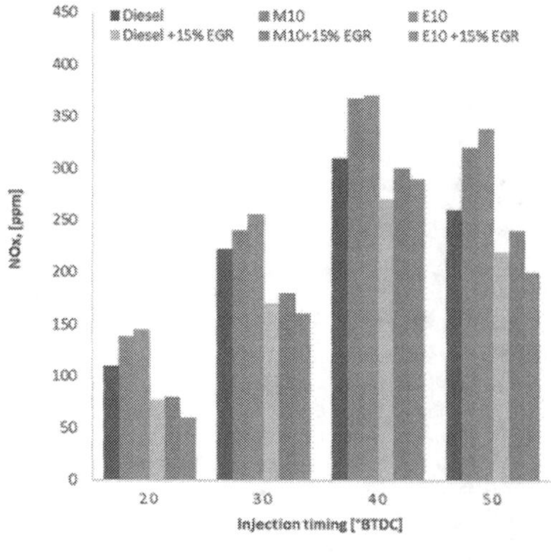

Fig. 8. Influence of E10, M10 and EGR on NOx under various injection timing.

Figure 8 shows that with a decrease in injection timing, the NOx emissions decreases. As a result, the entire combustion process moves along the line of expansion. As a result, fewer combustion products are exposed to the high temperature, resulting in less NOx emissions. Reducing injection timing combined with EGR is a good combination to reduce NOx formation as seen in Figure 8.

4 Conclusions

The effect of diesel/alcohol blends and EGR on engine characteristics were studied. The conclusions can be summarized as follows:

1. BSFC increases with increasing C_2H_5OH and CH_3OH in the fuel mixtures. A significant increase in fuel consumption was also observed for the fuel mixtures combined with a 10% EGR;
2. The obtained results show that when using EGR technology, an increase in HC and CO emissions is observed for all used fuels and fuel mixtures. The CO content is higher in fuels that do not contain oxygen in their molecular structure. Lower HC emissions are observed with alcoholic blends. This effect is most pronounced with C_2H_5OH-diesel blends;
3. NOx emission increases with increasing alcohol content in the fuel mixture. The increase in NOx is more significant in mixtures with CH_3OH. The obtained results show that as the load increases, NOx emissions also increase. NOx reduction with EGR at high loads conditions are greater compared to no load;
4. With a decrease in injection timing, the formation of NOx emissions decreases. Reducing injection timing combined with EGR is a good combination to reduce NOx formation.

References

1. T. Boden, G. Marland, R. J. Andres, Global, regional, and national fossil-fuel CO_2 emissions. Oak Ridge, Tenn, USA: Carbon Dioxide Inf Anal Center, Oak Ridge Natl Lab US Dep Energy; (2016)
2. J. Sheehan, V. Cambreco, J. Duffield, M. Garboski, H. Shapouri, An overview of biodiesel and petroleum diesel lifecycles. A report by US Department of Agriculture and Energy, p.1–35 (1998)
3. S. Saravanan, G. Nagarajan, S. Sampath, Combined effect of injection timing, EGR and injection pressure in NOx control of a stationary diesel engine fuelled with crude rice bran oil methyl ester, Fuel, Volume **104**, Pages 409-416, (2013)
4. M. Anandan, S. Sampatha, N. M. Sudharsanb, Effect of compression ratio and exhaust gas recirculation (EGR) on combustion, emission and performance of DI diesel engine with biodiesel blends. Global Journal of Research In Engineering 14 (**7**), 1-16, (2014)
5. Y. Çelebi, H. Aydın, Investigation of the effects of butanol addition on safflower biodiesel usage as fuel in a generator diesel engine. Fuel, **222**:385–93, (2018)
6. J. Vancoillie, J. Demuynck, L. Sileghem, M. Van De Ginste, S. Verhelst, L. Brabant, et al. The potential of methanol as a fuel for flex-fuel and dedicated spark-ignition engines. Appl Energy **102**:140–9, (2013)
7. M. Chaichan, Air pollutants and performance characteristics of ethanol-diesel blends in CI engines Engineering & Technology Journal, vol. **28**, No. 21, pp. 6365-6383 (2010)
8. L. Xing-cai, Y. Jian-guang, Z. Wu-gao, H. Zhen Effect of cetane number improver on heat release rate and emissions of high-speed diesel engine fueled with ethanol–diesel blend fuel. Fuel, **83**, 2013–20 (2004)
9. H. Bayraktar, An experimental study on the performance parameters of an experimental CI engine fueled with diesel–methanol–dodecanol blends. Fuel, **87**,158–64 (2008)

 https://doi.org/10.1051/e3sconf/202340402006

10. K. Murat, G. Ergen, M. Hosoz, Effects of the blends containing low ratios of alternative fuels on the performance and emission characteristics of a diesel engine, Fuel, Volume **112**, Pages 537-541, (2013)
11. B. Rossomando, I. Arsie, C. Pianese, Analysis of the impact of Diesel-Ethanol fuel blends on CI engine performance and emissions via multi-zone combustion modelling, Energy Procedia, Volume **126**, Pages 1059-1066, (2017)
12. A. Oliveira, A. M. Morais, O. S. Valente, J. R. Sodré, Combustion characteristics, performance and emissions from a diesel power generator fuelled by B7-ethanol blends. Fuel Process Technol **139**:67–72, (2015)
13. Z. H. Huang, H. B. Lu D. M. Jiang, K. Zengm, B. Liu, J. Q. Zhang, et al. Performance and emissions of a compression ignition engine fueled with diesel/oxygenate blends for various fuel delivery advance angles. Energy and Fuel **19**:403–10 (2005)

E3S Web of Conferences **404**, 02007 (2023) https://doi.org/10.1051/e3sconf/202340402007
EEPES 2023

Hybrid speed bump-PV-battery-grid power generation in a smart grid environment

Oluwafemi Oni, *Robert Thokozani* Skhosanaa, and *Omowunmi Mary* Longe*

Department of Electrical & Electronic Engineering Science, University of Johannesburg. Johannesburg, South Africa

Abstract. To meet the growing demand for clean energy and improved energy availability factor in South Africa, various solutions are being proposed from academia and industry. Therefore, in this study, a non-conventional method of generating clean energy was designed and constructed. The proposed system harnesses the kinetic energy of vehicles driving over speed bumps and converts it to electrical energy using a generator. This energy is supplemented with the energy generated from solar PV panels to provide a sustainable power supply for traffic and streetlights. The system is further designed to operate with a grid connection as a backup. An interface was further created to give real-time monitoring of electric power generation and consumption. The setup is also suitable for areas with limited/no access to electricity, and in the case of loadshedding. The results obtained during the testing of this prototype demonstrate the effectiveness of the system to provide a reliable source of electricity for street and traffic lights to ensure safety on the roads and avoidance of traffic congestion during loadshedding. The prototype is scalable to meet future power demands and feed into the grid in a smart grid environment.

1 Introduction

Energy crisis is becoming a serious problem in South Africa as a result of a shortage in the generating capacity and 58% energy availability factor (EAF) from the utility provider [1]. Besides, energy consumption has grown considerably due to the increasing population and additional loads [2]. The country's energy consumption is largely dependent on non-renewable sources such as natural gas, coal, and oil [3], which have caused significant atmospheric and environmental pollution, resulting in phenomena such as climate change, and other greenhouse gas effects. Given the current global economic uncertainty and the fluctuation of energy costs as a result of limited resources, it is critical to prioritize the usage of renewable energy sources for highly important tasks in our day-to-day living.

Emerging solution focuses on decentralized renewable energy (DRE) power production systems [4]. This can play an essential role in combating the low EAF in South Africa's urban and rural areas. Main reason because the decentralized solutions can be more cost-

* Corresponding author: omowunmil@uj.ac.za

effective, and highly reliable than the central grid for delivering power to remote, low-density or urban areas. Battery storage with a renewable energy source can further be incorporated to build a hybrid DRE power supply system. In comparison to the utility grid or single sourced energy systems, hybrid DRE systems have numerous advantages, including increased power supply reliability, environmental damage mitigation, smaller component sizes, and possibly lower unit energy costs. Solar-photovoltaic-based hybrid systems are a viable option in many South African cities since solar radiation is abundant in most of the provinces.

With the fast urbanization and increase in the number of internal combustion engine cars, urban transportation is a major contributor to increase in energy consumption, triggering a series of problems such as unclean energy consumption, high CO_2 emissions and air pollution [5]. Other facilities needed for a well mapped out urban centre are a well-lit streetlight, and a functional traffic controlling lights. However, a non-functioning streetlights and traffic lights in various regions have been cumulated to a rise in crime rates and accidents, particularly affecting pedestrians, including workers and students. Streetlights and traffic lights serve to enhance road safety by illuminating the streets and providing a right of pass, thereby rendering individuals more conscious of their environment. Statistics reveal that absence of adequate street and traffic lighting elevates the likelihood of robbery and vehicular accidents [6]. The use of renewable energy technology has shown to be a long-term answer to these transportation industry's problems.

Traffic is one of the most energy-intensive parts of human activities in the urban environment, and it is also one of the most wasteful. It was noted that just 30 percent of the fuel capacity is used to transfer kinetic energy to the vehicle during traffic, and most of this energy is lost in decelerating stages by braking. With over 30% of the population utilizing automobiles for daily activities, one possible avenue for alternate energy generation is the conversion of road friction generated by traffic calming devices such as speed breakers, speed humps, cushions, and speed charts. These vertical deflection devices are commonly used to reduce the amount of motor vehicle traffic and improve safety conditions by imposing a low-speed restriction of less than 40 km/h. These devices are needed mostly at decelerating locations, for example, urban road and pedestrian crossings, and important road exits because it enables the best possible mix of increased road safety and compensation of carbon emissions.

Recent research studies have attempted to identify new environmentally friendly energy sources that are both profitable and efficient. Roadway speed bumps are promising candidates and have been researched upon because of their constant exposure to diverse energy sources for energy gathering resources [7-10]. As a result, there is a growing interest in energy harvesting by collecting squandered unconventional sources and transforming them to a usable energy form. PV solar radiation wind hybrid power generation was used in [11] for street lighting, and heat energy harvesting from pavements was proposed in [12]. The aforementioned sources of energy are dissipated into the pavement as strain and thermal energy. These technologies were proposed for the roadside infrastructure network and to feed smart transportation systems that control traffic and health monitoring systems, electrical signs and lighting, etc. The decrease in different electronics circuitry size and power consumption due to recent advances in complementary metal-oxide semiconductors has further motivated researchers to develop new energy harvesting devices, such as electrohydrodynamic (EHD) devices that can be used to generate power from a variety of sources, including vibrations, air flow, and even human motion.

Therefore, in this paper, an embedded design of a speed bump and PV system were proposed for power generation for streetlights and traffic lights. The study also included a battery backup with availability for grid input. The result from the experimental testing of the design were recorded. The main contribution comes in the design of the hybrid DRE and implementation of internet of things (IoT) database using ThingSpeak for online monitoring of the design. The final design prototype was built to generate a minimal amount of power of up to 10 watts. This power could be used to power traffic lights, streetlights, or even small electronic devices. And thus demonstrate the feasibility of harnessing energy from road speed breakers and incorporating it into daily usage for traffic lighting and streetlight.

2 The photovoltaic solar cell current-voltage characteristic curve

The current-voltage (I-V) characteristic curves of a solar cell demonstrate the current and voltage properties of a given PV cell, module, or array, by providing its electrical features. This data gives a thorough examination of the cell's efficiency and capacity to convert solar energy. Understanding the I-V properties of a solar panel or cell, particularly P_{max}, is critical for determining its performance and solar efficiency. The PV solar cells convert the sun's radiant radiation directly into electrical energy. Solar energy conversion has become an increasingly essential element of the mix of renewable energy sources because of the growing need for clean energy and the sun's capability as a renewable energy source. As a result, consumer interest in efficient solar cells that convert sunlight directly into power is increasing more rapidly than ever. Because solar cells generate direct current (DC) electricity, we can create I-V curves for solar cells that represent the current versus voltage for a photovoltaic device [13]. The current-voltage (I-V) properties of a typical silicon PV cell is shown in Figure 1 during normal operating conditions [14]. To calculate the power output of a solar panel or cell, one needs to multiply its output current and voltage (IV). By performing this multiplication for each voltage point between short-circuit and open-circuit conditions, a power curve can be obtained for a specific radiation level [15].

A PV cell's I-V curve is the superposition of the dark I-V curve of the solar cell diode and the light-generated current. The light causes the I-V curve to drop down towards the fourth quadrant, where power may be recovered from the diode. Illuminating a cell increases the typical dark currents in the diode, causing the diode law to become Equation 1, which when rearrange will give the voltage relationship shown in Equation 2.

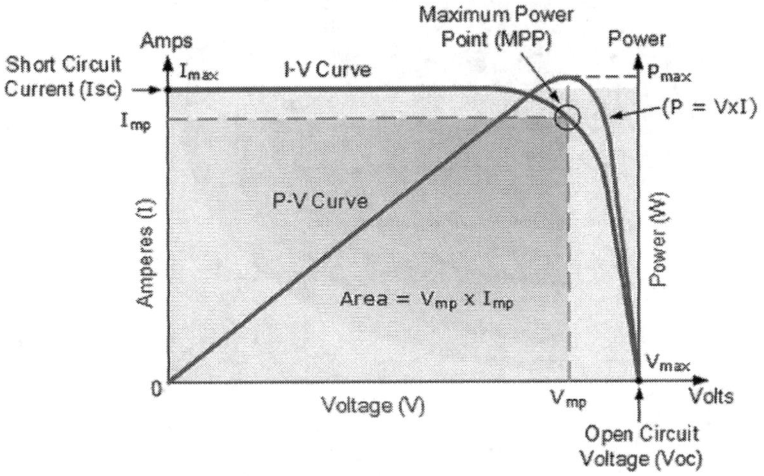

Fig. 1. A solar cell I-V characteristics curve [15].

$$I = I_0[\exp\left(\frac{qV}{nkT}\right) - 1] - I_L; \tag{1}$$

$$V = \frac{nkT}{q} ln(\frac{I_L - I}{I_0}). \quad (2)$$

where I_L = light generated current.

The relationship between the output current and voltage of a photovoltaic array summarizes its electrical properties. The quantity and strength of solar insolation, also known as solar irradiance, determines the level of output current (I), while the operational temperature of the solar cells impacts the output voltage (V) of the cell array.

3 Roller design model

This study proposed a methodology for generating electricity by utilizing a roller as shown in Figure 2 [7] placed between a speed bump and a specialized grip on the speed breaker. When a vehicle drives over the speed breaker, the roller is rotated, thereby turning the shaft of a generator to produce electrical energy. Mathematical formulas were employed in this study to determine the amount of electricity generated. Firstly, the simple circuits of the generator with the internal resistance and inductance (R_i and L_i) are respective shown in Figure 3. The load resistor, which is the point of harvesting the generated power is also shown and denoted as R_L. This diagram denotes an ideal generator, i.e., with a negligible value of L_i. According to the electromagnetic model of the generator in Figure 3, the electromotive voltage and electromagnetic torque is shown in Equations 3 and 4 respectively. Other parameters are the induced electromotive voltage V_{gen}, back electromotive voltage constant K_{gen}, rotational speed of the generator ω_{gen}, torque constant K_T, and external current i_{gen}. At low frequency of f, the internal inductance L of the geneator is usually small, that is, $FL< (R_i + R_L)/2\pi$. Therefore, the internal inductance L can be negligible. The final torque of the generator is given in equation 5 [16].

Fig. 2. Prototype of a speed bump design for energy recovery [7].

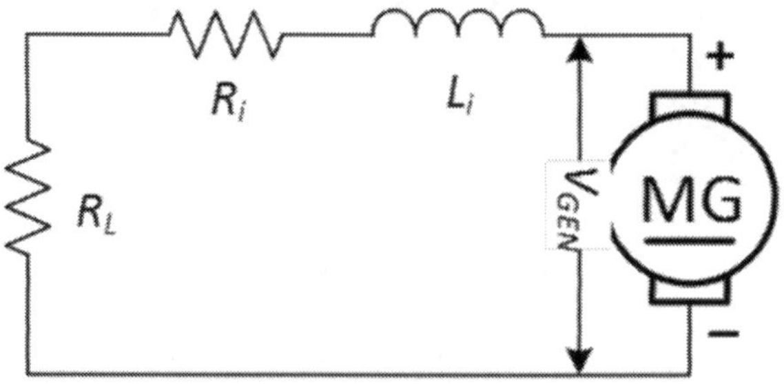

Fig. 3. Simplified model of a dc generator for the speed bump converter.

$$V_{gen} = K_{gen}\omega_{gen}; \quad (3)$$

$$T_{gen} = K_T i_{gen}; \tag{4}$$

$$T_{gen} = K_T \frac{V_{gen}}{R_i + R_L} = \frac{K_T K_{gen} \omega_{gen}}{R_i + R_L}. \tag{5}$$

This study offered a method in which a roller is placed between a speed bump and a grip is provided on the speed breaker such that when a vehicle drives over the speed breaker, the roller is turned. The generator produces electricity when the generator's shaft spins. The torque power relationship is given in equation 6. The angular speed of the generator equation is given in equation 7 while equation 8 gives the power equation.

$$\tau = P\omega; \tag{6}$$

$$\omega = \frac{2\pi N}{60}; \tag{7}$$

$$P = VI \quad , \tag{8}$$

where: P is the power generated (output power), ω is the angular speed of the motors, τ is the torque of the motors. N is the rotational speed of the motors, V is voltage produced by the motors, and I is the current computed from each subsystem.

4 Design prototype

The diagram in Figure 4 shows how each subsystem was integrated with the system. It consists of the speed bump arrangements, different relays, solar PV, grid electricity and different peripherals working concurrently to provide an efficient system.

The speed breaker is a spring that is supported at both ends, which can create linear motion. Pistons are placed beneath the speed breaker, and when a vehicle drives over it, the pistons are pushed down. This movement allows the piston to compress the oil beneath it, which is then transported to an accumulator. The accumulator is connected to a motor that generates torque from the compressed oil. As the roller turns the shaft of the DC motor, the torque is converted into electrical energy. Therefore, the kinetic energy produced by the motor's spinning is transformed into electrical energy, which is then stored in a battery.

LDR (light dependent resistor) was used to make sure that lights go on at a certain time. A timer switch is required to achieve this through the use of a timer switch.

The system is designed to made used of three sources of power namely, roller speed bump power generated via kinetic energy harvesting, solar PV, and the grid. The system is also equipped with battery bank. The Arduino used has been programmed with order of preference in selecting the sources. The relays help in the switching between these sources. Power generated via speed bump and the Solar PV are the main source of power for this system. They directly charge the battery, and thus further supply the street and traffic lights. The grid serves as a backup in a rare case of when there is no supply of power through either of the primary sources.

According to technical sources, it has been reported that the ESP8266 module has the capability to act as a host for an application or alternatively, it can offload all Wi-Fi IEEE 802.11 standard networking functions from another application processor. It has been noted that every ESP8266 module is pre-programmed with an AT command set firmware, which implies that when connected to an Arduino board, it offers similar Wi-Fi capabilities as a Wi-Fi Shield. Furthermore, the ESP8266 module is recognized as a highly cost-effective board and has a large and continuously expanding community of users.

The most important component is the Arduino MEGA because it is the one that controls every subsystem. Arduino mega 2560 serial monitor results from the grid, solar, battery, and roller. Results generated through this monitoring are shown in Figure 5. Each subsystem is represented by its voltage, current and power. This Serial Monitor results show the real-time current, voltage, and power of the different subsystems. The first value is current, the middle value is voltage, and the final value is power. The battery stays the same for most part of the testing because the roller and solar take time to charge the battery. The values that change a lot throughout the experiment are for the roller because cars pass over the roller at different times, with different speeds and weights. The polycrystalline solar panels used are two 1.5 V 0.25 W with 14%-16% efficiency placed on top of the streetlights.

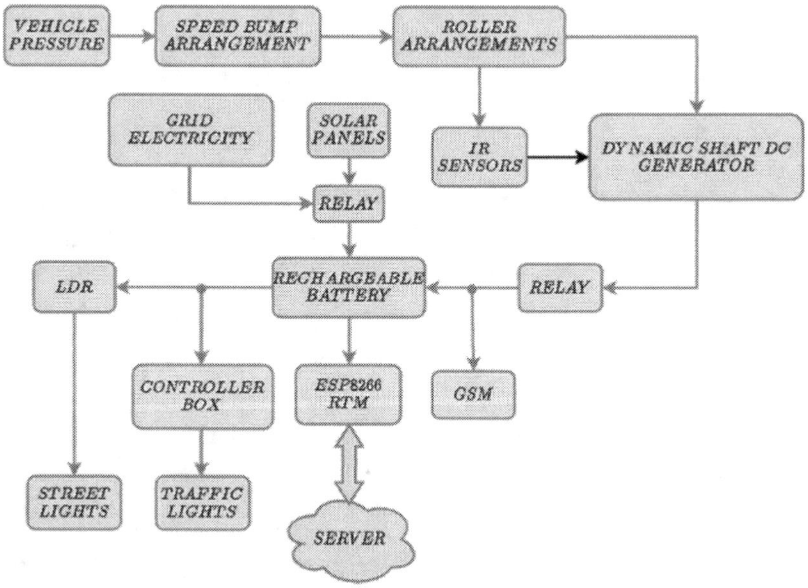

Fig. 4. Design of database table.

Roller,0.21,1.69,0.36	Roller,0.24,1.69,0.40
Solar,0.29,0.44,0.13	Solar,0.24,0.44,0.10
Grid,0.29,0.47,0.14	Grid,0.32,0.47,0.15
Battery,0.26,12.39,3.27	Battery,0.32,12.39,3.92
Roller,0.24,1.69,0.40	Roller,0.21,1.69,0.36
Solar,0.26,0.44,0.12	Solar,0.32,0.44,0.14
Grid,0.32,0.47,0.15	Grid,0.29,0.47,0.14
Battery,0.29,12.39,3.60	Battery,0.32,12.39,3.92
Roller,0.21,1.69,0.36	Roller,0.21,1.69,0.36
Solar,0.26,0.44,0.12	Solar,0.29,0.44,0.13
Grid,0.26,0.47,0.12	Grid,0.34,0.48,0.16
Battery,0.32,12.39,3.92	Battery,0.34,12.39,4.25
Roller,0.26,1.69,0.45	Roller,0.18,1.69,0.31
Solar,0.29,0.44,0.13	Solar,0.29,0.44,0.13
Grid,0.32,0.47,0.15	Grid,0.34,0.47,0.16
Battery,0.29,12.39,3.60	Battery,0.32,12.39,3.93
Roller,0.21,1.69,0.36	Roller,0.18,1.69,0.31
Solar,0.26,0.44,0.12	Solar,0.32,0.44,0.14
Grid,0.29,0.47,0.14	Grid,0.34,0.47,0.16
Battery,0.32,12.39,3.92	Battery,0.32,12.39,3.92

Fig. 5. Serial monitor results.

5 Testing and results

Figure 6 shows the results for the roller Speed Bump Roller mechanism from the serial monitor. A minimum of four results were selected. Each time a car passes over the ramp, the current and voltage are recorded and then the power is computed from the product of the two values.

The IoTs aspect of this research involves the use of web-based tools to monitor the data generated from the implementation of this design. The website displays the different results of the subsystems. On the left tab of the website, all the components displayed are from the roller, solar and Grid, the battery Bank, notifications, and cost. Clicking on one of the tabs will display the required results of that specific system. For the solar, the amount of power generated from the sun is going to be shown and when the grid electricity is on, the results related to the grid will be shown. Notifications happen when there is a fault in the system and display a notification when the energy in the battery is depleted. This part of the experiment uses the GSM SIM 800L to send notifications to the user when there is a fault in the system as shown in Appendix B and C. The user needs to be notified through a fast-messaging system and the GSM 800C was used to send messages to the user when there is a fault in the system e.g., when there is no current flowing in the circuit a message will be sent, which as well signifies a fault in the system.

The following results were taken using ThingSpeak during the testing of all the different subsystems. The first results, shown in Figures 7a and 7b, is for the roller speed bump. The data were taken in July and the second results were taken in October, the straight line in between these dates does not display a trend, it is a space in time where no results were taken. The results displayed are only for one reading, although multiple readings were taken which increases the reliability of the data taken from experiments.

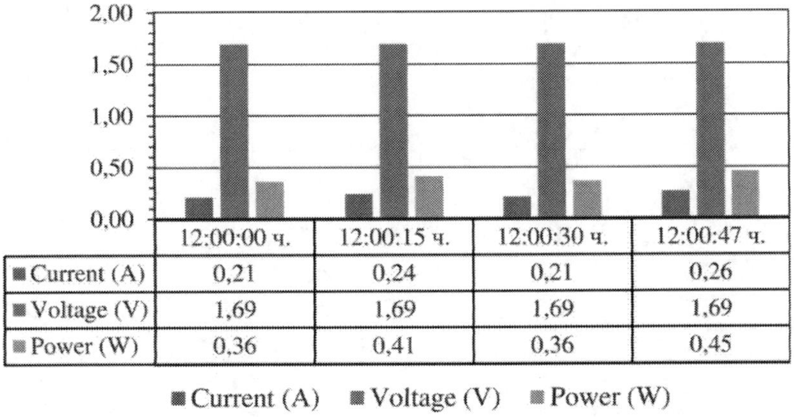

	12:00:00 ч.	12:00:15 ч.	12:00:30 ч.	12:00:47 ч.
■ Current (A)	0,21	0,24	0,21	0,26
■ Voltage (V)	1,69	1,69	1,69	1,69
■ Power (W)	0,36	0,41	0,36	0,45

Fig. 6. Roller Mechanism result showing the results (V-I-P vs Time) results from serial monitor.

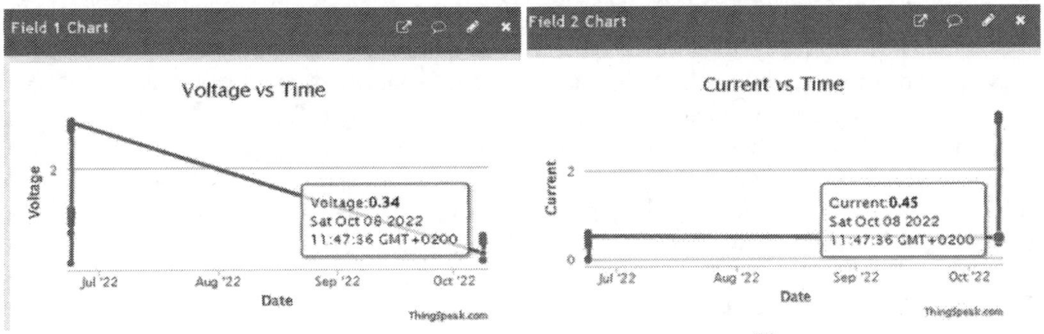

(a) (b)

Fig. 7. Roller speed bump ThingSpeak showing (a) voltage and (b) current results.

During the usage of Solar PV, data were taken based on the reading on the system display for voltage and current at different time interval as shown in Figure 8 from experiments. The value taken was low because of the size of the polycrystalline Solar panel that was used. As this project is scalable, increasing the power output is possible with the increase in the sizing of the solar panel. Further results were taken using ThingSpeak during the testing of all the different subsystems when Solar PV system was connected. The data were taken in October at different time interval as shown in Figures 9a and 9b. The line in between also signifies dates and time where no results were taken. The results displayed are only for one reading, although multiple readings were taken which increases the reliability of the data taken from experiments.

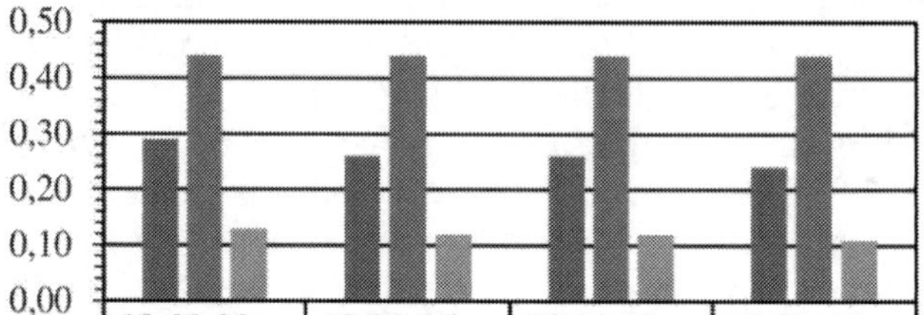

	12:01:00 ч.	12:01:15 ч.	12:01:30 ч.	12:01:45 ч.
Current (A)	0,29	0,26	0,26	0,24
Voltage (V)	0,44	0,44	0,44	0,44
Power (W)	0,13	0,12	0,12	0,11

Current (A) Voltage (V) Power (W)

Fig. 8. Solar PV results from serial monitor (V-I-P vs Time).

Figure 10 shows the results for the grid electricity from the serial monitor. A minimum of four results were selected. The current and voltage are recorded and then the power is computed from the product of the two values.

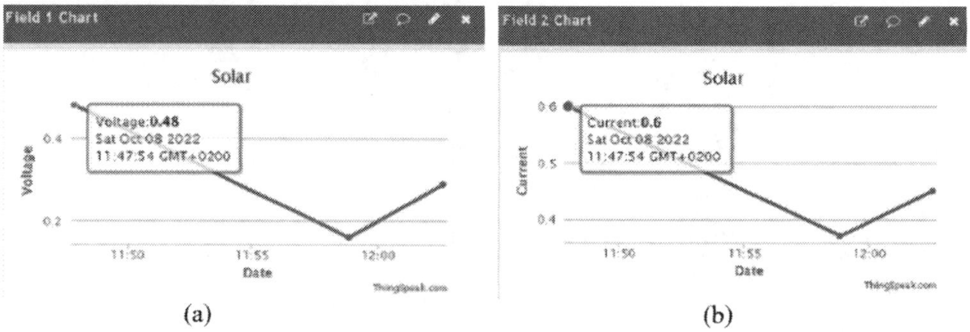

Fig. 9. Solar PV (a) voltage and (b) current vs time results.

Figure 11 shows the results for the battery bank from the serial monitor. The current and voltage are recorded and then the power is computed from the product of the two values. A minimum of four results were selected and the results show that the battery was being used since the power values increased after it was charged by the roller and solar energy. The exterior and interior of the prototype hybrid power generator are presented in Figures 12a and 12b. The figure includes components used in this prototype are: Rechargeable Battery, Arduino MEGA 2560, Polycrystalline Solar Panel, LED Lights, IR Sensors, GSM 800L, Current Sensors, Relays and boost converters, Breadboard, and connecting cables.

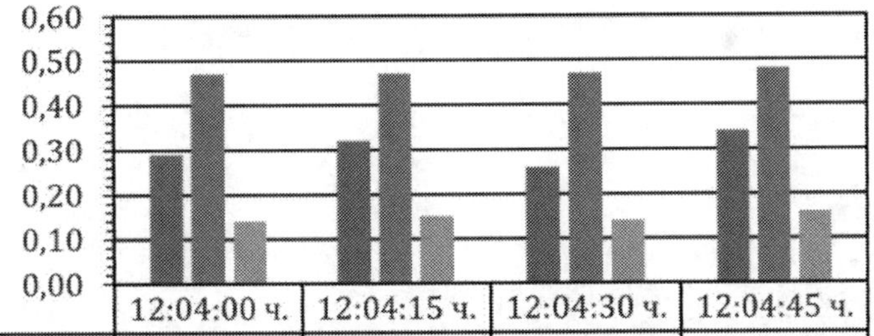

	12:04:00 ч.	12:04:15 ч.	12:04:30 ч.	12:04:45 ч.
Current (A)	0,29	0,32	0,26	0,34
Voltage (V)	0,47	0,47	0,47	0,48
Power (W)	0,14	0,15	0,14	0,16

Current (A) Voltage (V) Power (W)

Fig. 10. Graph showing results for grid electricity (V-I-P vs Time).

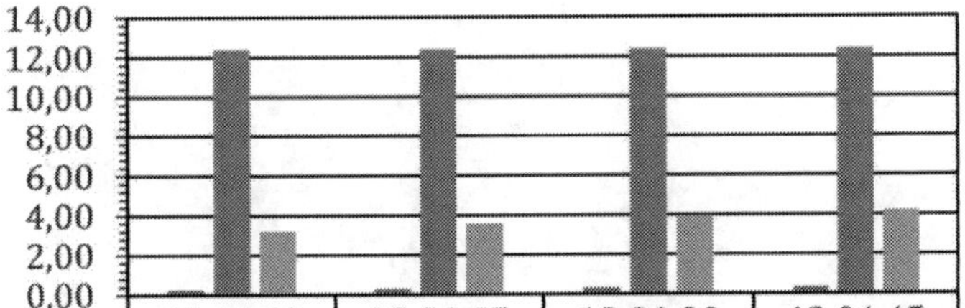

	12:04:00 ч.	12:04:15 ч.	12:04:30 ч.	12:04:45 ч.
Current (A)	0,26	0,29	0,32	0,34
Voltage (V)	12,39	12,39	12,39	12,39
Power (W)	3,22	3,59	3,96	4,21

Current (A) Voltage (V) Power (W)

Fig. 11. Graph showing results for the battery bank (V-I-P vs Time).

The testing of the prototype using a toy car is shown on Figure 13. As soon as the truck passes the 1st IR Sensor, it will detect the truck and cause the roller to move, and it stops when the car has passed over the 2nd IR sensor. The speed at which the car passes the roller speed bump determines the amount of power generated at the outside end of the generator. The output, being a small prototype, is made to be in dc since all the needed loads are in dc. This way, losses are minimized because if converted to AC, the losses due to the IGBTS or diode switches will lead to a reduction in the output power for the streetlight and traffic light load. The results presented in this work contribute to the fact that hybrid energy generation would play a big role in the energy generation for the future smart grid both in rural and urban areas as earlier asserted by [7, 8, 11, 12, 17, 18]. This work on a large scale can feed into the grid to increase grid reliability and contribute to improved EAF in the country and reduce hours of loadshedding in the areas where hybrid speed bump-PV-Battery-Grid systems are in the country. The deployment of this solution could contribute to the Minister of Electricity's goal of achieving 75% EAF [19].

The results generated during the IoTs testing of this prototype are shown in the Appendix section. The ThingSpeak website used is shown in Appendix A. This diagram shows different subsystem as combined. And being equipped with fault tracking and report capabilities, the diagram in Appendix B shows the notification which is directly sent to the user via SMS. The diagram in Appendix C shows the exact location of faults or report on the usage of the battery.

a)

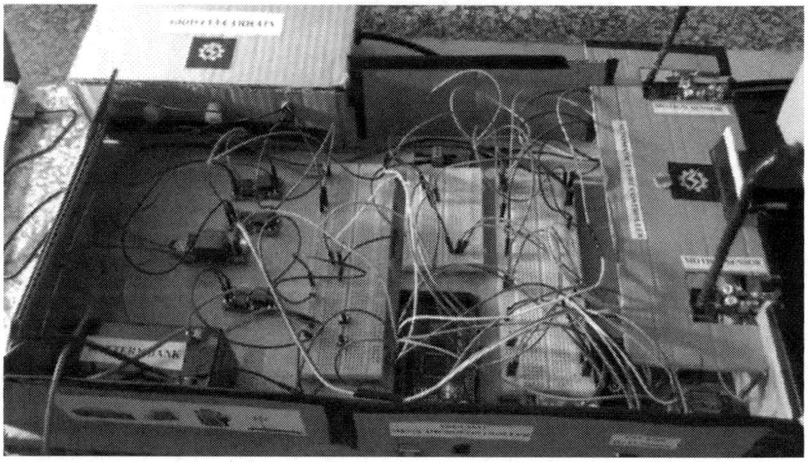

b)

Fig 12 Exterior (a) and interior (b) of the designed solar PV-speed bump hybrid power generator prototype.

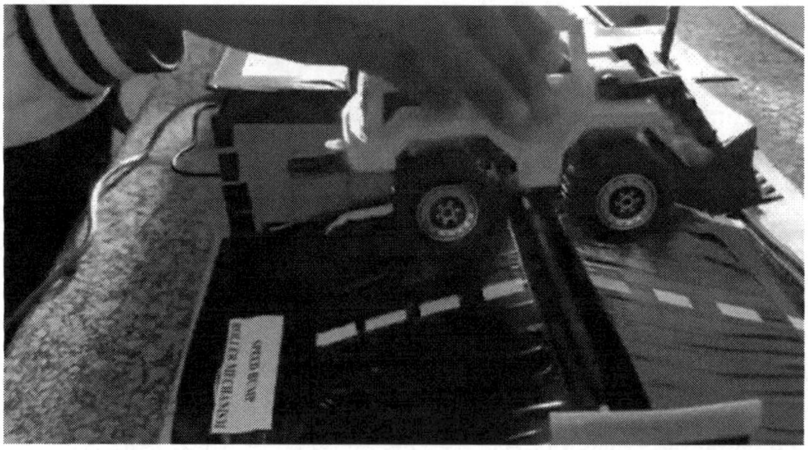

Fig. 13. A toy car going over the speed bump.

The prototype costed about R2400 (approximate $125) as at the time of writing this paper. Large scale production of the device would reduce the per unit cost of the system with good return-on-investment for the company that would manufacture them.

6 CONCLUSION

In this paper, an embedded design of a speed bump-PV-Battery-Grid clean power generation were proposed for streetlights and traffic lights. This system is designed to recover energy from traffic on speed bumps. The usefulness of this design is to help alleviate over-reliance on the grid, increase security at night on the roads and improve grid EAF from its present state of 58%. The performance of the designed prototype has been analyzed by carrying out a serial monitoring of the device using ThingSpeak. ThingSpeak is a web base tool link with the design to give a real time data capturing of the device to mimic a smart grid environment. The test carried out confirmed that the prototype, which was built to supply power for street and traffic lights, can also be applied for monitoring traffic or highway indication systems, and for infrastructure utilities such as electrical signage, lighting etc. The practical implementation of this design on South African roads will come in handy in reducing the traffic congestion during loadshedding and save time taken by commuters to get to their destinations during loadshedding. It will also play a crucial role in augmenting security at night in our communities and improving the quality of life by artificially prolonging the duration of daylight hours, thereby facilitating extended hours for commercial activities and improved economy.

Some of the limitations of the study that can be mitigated in future designs include use of monocrystalline solar panels instead of polycrystalline solar panels, and use of at least a raspberry-pi instead of an Arduino Uno microcontroller. Also, more budget should be given to practically deploy the proposed solution on our roads.

References

1. O. Analytica, "Eskom bailout will not solve South Africa power crisis," Emerald Expert Briefings, no. oxan-db, (2023).
2. M. C. Udeagha, N. Ngepah, Econ. Change and R., **55**, 3, pp. 1767-1814, (2022).
3. B. Büscher, S. Koot, L. Thakholi, Envrnmt and Planning E: Nat. and Space, p. 25, (2022).
4. V. Harish, N. Anwer, and A. Kumar, Sust. Energy Tech and Assmnts, **52**, p. 102032, (2022).
5. K. A. Collett, S. A. Hirmer, H. Dalkmann, C. Crozier, Y. Mulugetta, M. D. McCulloch, Energy Strat. Reviews, **38**, p. 100722, (2021).
6. V. N. Sumantri, A. I. Rifai, F. Ferial, Citizen: Jurnal Ilmiah Multidis. Indnsia, **2**, no. 5, pp. 703-711, (2022).
7. A. Pirisi, M. Mussetta, F. Grimaccia, R. E. Zich, IEEE Trans. on Int. Trans. Sys., **14**, no. 4, pp. 1983-1991, 2013.
8. I. R. R. KranthiMadala, M. F. Shaik, Recnt Dev in Electr and Coms Sys, **32**, pp. 475-483, (2023).
9. M. Gholikhani, R. Nasouri, S. A. Tahami, S. Legette, S. Dessouky, A. Montoya, Applied Energy, **250**, pp. 503-511, (2019).
10. M. Sun, W. Wang, P. Zheng, D. Luo, Z. Zhang, Sensors and Actuators A: Phy, **323**, p. 112648, (2021).

11. W. R. Nyemba, S. Chinguwa, I. Mushanguri, and C. Mbohwa, Procedia Manufac. **35**, pp. 285-290, (2019).
12. A. Dawson, R. Mallick, A. G. Hernandez, P. K. Dehdezi, Climate Change, Energy, Sust. and Pavements, pp. 481-517, (2014).
13. A. Orioli, Renewable Energy, **145**, pp. 725-743, (2020).
14. J. M. Greulich et al., Solar Energy Mat. and Solar Cells, vol. 248, p. 111931, (2022).
15. S. A. Afghan, H. Almusawi, and H. Geza, "Simulating the electrical characteristics of a photovoltaic cell based on a single-diode equivalent circuit model," in MATEC Web of Conferences, 2017, **126**: EDP Sciences, p. 03002. (2017)
16. L. Wang, P. Todaria, A. Pandey, J. O'Connor, B. Chernow, L. Zuo, "An electromagnetic speed bump energy harvester and its interactions with vehicles," IEEE/ASME Trans. on Mech., **21**, no. 4, pp. 1985-1994, (2016).
17. L. Olatomiwa, A. A. Sadiq, O. M. Longe, J. G. Ambafi, K. E. Jack, T. A. Abd'azeez, S. Adeniyi, Energies, vol. 15, no. 9554, (2022).
18. R. Gilbert, O. M. Longe, "A Feasibility Study on Optimal RES Microgrid Design for Rand West Municipality," in IEEE Power Africa Conference (PAC), Nairobi, Kenya, 2020, pp. 1-5. (2020)

C. Parton, " New Eskom board's 75% energy availability factor mandate is impossible, says Mark Swilling," News24, 24 October 2022. Available at: https://www.news24.com/fin24/economy/new-eskom-boards-75-energy-availability-factor-mandate-is-impossible-dbsa-chair-20221024. Accessed on 29 April 2023. (2023)

APPENDICES

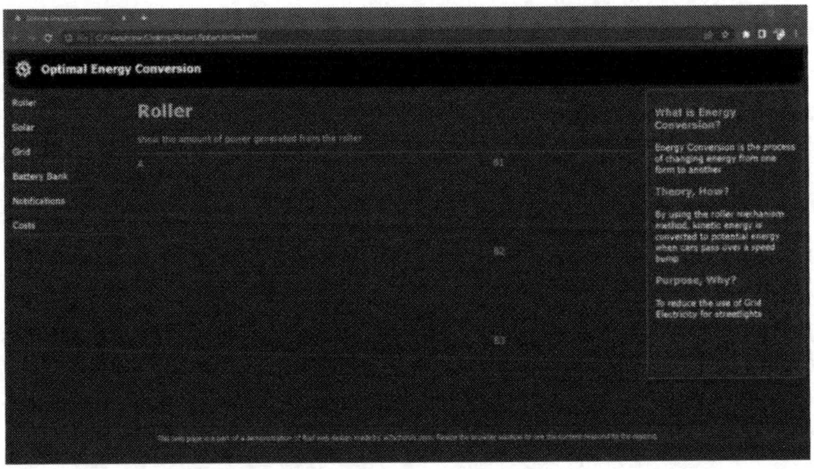

Appendix A. Website for ThingSpeak Results

Appendix B. Serial Communication Notification via SMS.

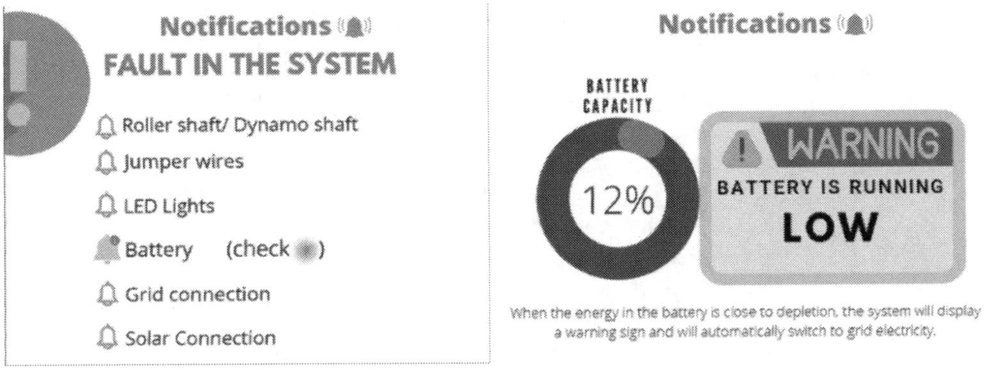

Appendix C. Fault in the system notification.

The optimized power flow control system for the photovoltaic DC microgrid

Elvin Yusubov[1*], *Lala* Bekirova[2]

[1]Azerbaijan State Oil and Industry University, Instrumentation Engineering Department, AZ1010, Baku, Azerbaijan
[2]Azerbaijan State Oil and Industry University, Instrumentation Engineering Department, AZ1010, Baku, Azerbaijan

Abstract. The power flow control systems play a significant role in DC microgrids with photovoltaic inputs to supply the load with continuous power. The output power of the photovoltaic modules could experience a decline due to fluctuations in solar irradiation and temperature, which necessitates the use of batteries and the utility grid to reduce the negative effects of undesirable variations. However, an efficient control strategy is necessary to ensure an uninterrupted energy supply to the load units. This paper proposes an improved control of energy flow based on a State-of-Charge battery power estimation technique using the Coulomb counting method. By accurately estimating the available power from the batteries using the State-of-Charge technique, the microgrid is able to determine to assess if it requires to switch to the grid when the power output from photovoltaic modules is insufficient to meet the load demand. The proposed method also eliminates the need for DC bus voltage level-based approaches to charge or discharge the batteries with the advantages of the significant reduction in DC bus voltage variations. The simulation results of the proposed approach show that it provides satisfactory control performance to meet the load demand.

1 Introduction

DC microgrid energy storage and control systems have become increasingly important as a means of providing reliable and sustainable power supply in various applications, such as in remote areas, electric vehicles, and smart buildings. The DC microgrid comprises a network of interconnected power sources, storage devices, and loads that operate at low DC voltages [1, 2]. The simple structure of the DC microgrid is shown in Figure 1. The microgrid can operate independently or can be connected to the main power grid [3]. Energy storage and control systems play a crucial role in DC microgrids by ensuring the continuous power supply to the loads, regardless of changes in the power supply or demand [4-7]. The primary function of energy storage systems is to store excess power when it is available and release it when there is a shortage. Control systems, on the other hand, manage and coordinate the operation of the different components in the microgrid to ensure optimal performance and efficiency.

* Corresponding author: elvinyusifov05@gmail.com

https://doi.org/10.1051/e3sconf/202340403001

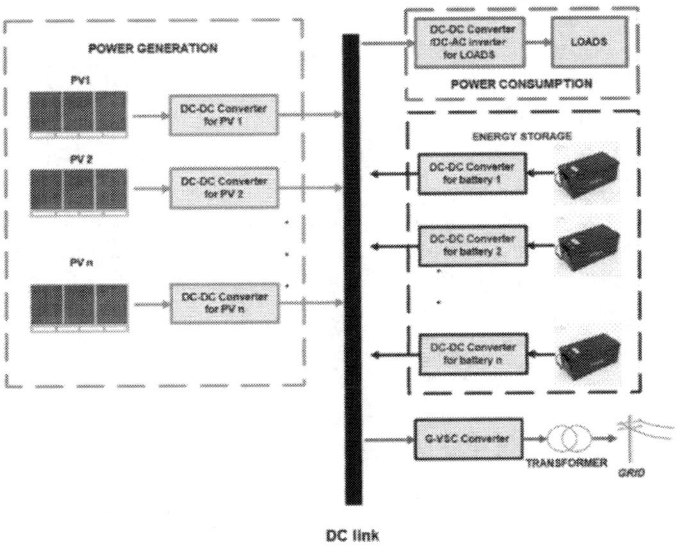

Fig. 1. Simple DC microgrid model.

The integration of renewable energy sources, such as solar PV and wind power, in DC microgrids has provided an opportunity to reduce carbon emissions and promote sustainable energy use. However, the variable nature of these energy sources presents a challenge in maintaining a stable power supply to the load [8, 9]. To mitigate this challenge, energy flow control systems are employed to balance the power supply and demand in the microgrid.

Several strategies are developed to achieve efficient power delivery to the load. In recent times, numerous authors have employed the Kalman filter (KF), a robust algorithm, for estimating the State of Charge (SoC). However, this approach faces a significant limitation in that a suitable battery model is required for the KF to operate effectively [10]. The energy management of the DC microgrid can be considered either in autonomous or grid-connected modes. Although the autonomous mode is useful where the grid power is not available, DC microgrids with utility grids can perform efficiently to compensate for the possible power fluctuations in PV modules and batteries. However, many existing grid-connected DC microgrids do not have an efficient switching strategy to the grid, since the grid power usage must be minimized to minimize energy cost [11, 12]. DC bus signalling methods are one of the prominent methods employed in battery management systems to control the charging and discharging of the battery units. In the DC bus signalling method, the DC bus voltage determines the charging or discharging of the battery units [13, 14]. Noise and interference susceptibility, low interoperability and compatibility, limited transmission distance and low-speed communication are the primary disadvantages of this method.

The primary aim of the paper is to develop an efficient power flow control system for DC microgrids to reduce the shortcomings of the above-mentioned methods.

2 System modelling

To create an efficient energy control system, the internal power flow structure of the DC microgrid must be understood.

2.1 PV modules to DC bus connection

To extract maximum power from the PV modules, MPPT trackers are employed. MPPT controllers are electronic devices used to optimize the power output of photovoltaic solar panels. The maximum power point (MPP) is the point at which a solar panel operates at its maximum power output. MPPT controllers track the MPP of the solar panel and adjust the voltage and current to maintain it at this point. It increases the efficiency of the solar panel, enabling it to produce more power. MPPT controllers are commonly used in both off-grid and grid-connected solar systems to charge batteries, power electrical loads, and feed excess power back into the grid. For higher efficiency, the Moth-Flame optimization-based MPPT controller is employed in the modelling of the DC microgrid [15].

2.2 Battery pack to DC bus connection

The process of connecting a battery pack to a DC bus involves establishing both electrical and physical connections between these two components. A battery pack consists of multiple battery cells linked together to provide a specific voltage and capacity, while the DC bus serves as a distribution system that connects various components within a system. To ensure a safe and efficient connection between the battery pack and the DC bus, a power electronics interface is utilized. This interface employs bidirectional converters, allowing energy to flow in two directions: from the battery pack to the DC bus and vice versa. In the modelling of DC microgrids, the bidirectional SEPIC-ZETA converter is commonly used as it enables the charging or discharging of the batteries without requiring a change in the DC bus voltage.

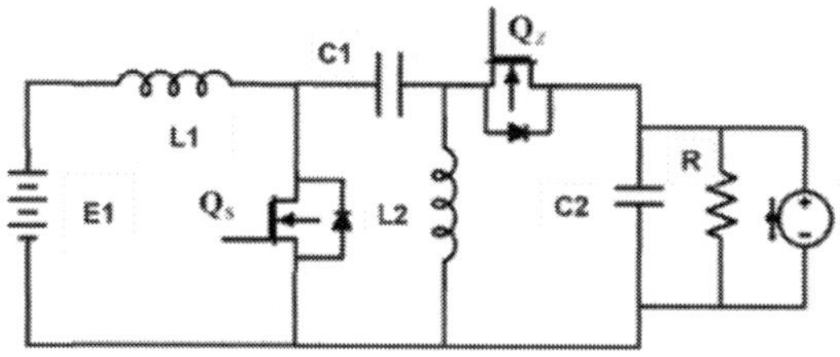

Fig. 2. Bidirectional SEPIC-ZETA converter.

The bidirectional SEPIC-ZETA converter, depicted in Figure 2, contains the following components:

• V_{E1} and V_{E2} are the battery voltage and DC bus voltage, respectively.

•C_1, C_2 are the capacitors,

•L_1, L_2 are the inductors,

• Q_s and Q_z are transistors,

• R is the load resistance.

The bi-directional SEPIC-ZETA DC-DC converters are equipped with two switching transistors that allow the power to flow in two directions. These converters can be used for battery charging and discharging modes in the DC-DC converter mode.

During battery discharge mode (mode 1), the bi-directional SEPIC-ZETA converter is utilized to transfer the stored energy from the batteries to the DC bus. In this mode, the converters act as a SEPIC converter by switching Q_Z transistor off, and activating Q_S transistor. The average dynamic model of the SEPIC converter is constructed with the following equations:

$$L_1\frac{di_{L1}}{dt} = V_{E1} - (1-D)(V_{C1}+V_{C2})\,; \tag{1}$$

$$L_2\frac{di_{L2}}{dt} = DV_{C1} - (1-D)V_{C2}\,; \tag{2}$$

$$C_1\frac{dV_{C1}}{dt} = (1-D)i_{L1} - Di_{L2}\,; \tag{3}$$

$$C_2\frac{dV_{C2}}{dt} = (1-D)(i_{L1}+i_{L2}) - \frac{V_{C2}}{R}, \tag{4}$$

where i_{L1} is the current through the inductor L_1, i_{L2} is the current through the inductor L_2, V_{C1} is the voltage across the capacitor C_1, V_{C2} is the voltage across the capacitor C_2, V_{E1} is the battery voltage, *R* is the load resistance, *D*-duty-cycle corresponding to the on the state of the transistor, *1-D* is the duty cycle corresponding to the off state of the transistor.

During battery charging mode, the bi-directional ZETA converter is activated and charges the batteries using the DC bus power. In this mode, the converter operates in ZETA mode, and transistor Q_z is activated, while transistor Q_s is closed. The average dynamic model of the ZETA converter is constructed with the following equations:

$$\frac{di_{L2}}{dt} = D\frac{V_{C2}}{L_2}+(1\text{-D})\frac{V_{C1}}{L_2}; \tag{5}$$

$$\frac{di_{L1}}{dt} = D\left(\frac{V_{C2}}{L_1} - \frac{V_{C1}}{L_1}\right) - \frac{V_{E1}}{L_1}; \tag{6}$$

$$\frac{dV_{C1}}{dt} = D\frac{i_{L1}}{C_1} - (1\text{-D})\frac{i_{L2}}{C_1}. \tag{7}$$

2.3 VSC Connection

The G-VSC (Grid-Connected Voltage Source Converter) is responsible for managing power transfer between the main grid and the DC microgrid. Its primary function is to regulate the voltage of the DC link to the desired level. When there is a power shortage in the network, the G-VSC controller generates signals to import power from the main grid. Conversely, in situations of excess power, the G-VSC exports power to the utility while ensuring that the DC link voltage remains stable. However, the maximum current capacity of the G-VSC is limited by the converter controls at its nominal value. This limitation restricts the G-VSC's ability to effectively regulate the DC bus voltage in the grid-connected mode of operation.

2.4 Battery SoC estimation

The State-of-Charge (SoC) of a battery refers to the current level of its capacity, represented as a percentage of its maximum capacity. It is a crucial metric used to measure the amount of energy stored in the battery relative to its full capacity. SoC plays a vital role in battery management systems as it helps to determine the remaining energy available and the rate at which the battery can be charged or discharged. The general mathematical formula for SoC is the following:

$$SoC(t)=\frac{Q_{actual}}{Q_{rated}}\,100\%, \tag{8}$$

where Q_{rated} and Q_{actual} are the capacity of the battery packs.

The Coulomb calculation method is based on the available battery capacity, the number of charges entering and leaving the battery [11].

$$SoC(t) = SoC(t_0) - \frac{1}{Q_{rated}} \int_{t_0}^{t_0+\tau} \eta I_b(t)dt, \quad (9)$$

where,
$SoC(t_0)$ is the initial SoC,
I_b- is battery current,
η_b efficiency of the battery (between 0 and 1).

The precision of the measurement method provided relies heavily on accurately determining the battery's current and initial state of charge (SoC). By obtaining the initial charge value, it becomes feasible to calculate the battery's SoC by integrating the charge and discharge currents over a specific timeframe. However, it's important to note that the amount of charge discharged during charging and discharging is consistently lower than the accumulated charge in the battery. This implies the presence of energy losses during the charging and discharging process, which are characterized by the efficiency coefficient.

The battery operates in three different modes: charging, discharging, and when there is no circuit connected, known as the open circuit mode.

In the charging mode, the battery's voltage and current are provided through constant sources of current and voltage, respectively. By maintaining a constant charging current, the battery's voltage gradually rises and approaches the specified threshold. When charging the battery in voltage mode, the charging current initially decreases rapidly and then at a slower rate. Eventually, when the battery reaches full charge, the current value becomes 0.

In the discharging mode, the voltage at the battery terminals gradually decreases over time. A higher current load leads to a faster decline in terminal voltage, resulting in a reduced duty cycle. By examining the relationship between the state of charge (SoC) and the discharge voltage at various current values, we can determine the initial SoC during the discharge period.

In open circuit mode, the battery is disconnected from the load circuit. The open circuit voltage can be used to determine SoC. To find the current from (9).

$$I_b(t) = -\frac{Q_{rated.}}{\eta} \frac{d}{dt} SoC(t). \quad (10)$$

The power can easily be calculated from (10).

$$P_{bat.}(t) = -\frac{U_b Q_{rated.}}{\eta} \frac{d}{dt} SoC(t). \quad (11)$$

3 Proposed control strategy of the energy storage system

3.1 Proposed Power Flow Control Method

DC microgrids receive the power from three primary types of the power sources. These power sources include the PV (photovoltaic) modules, battery packs and utility grid. The primary purpose of the power flow control is to deliver enough power to the load units. The proposed power flow control algorithm and structural diagram are given in Figure 3, and Figure 4, respectively. In the proposed algorithm, two types of switching states are assumed. When switching state is 1, the corresponding unit (PV, Battery or grid) is connected to the load unit. When the state is 0, the corresponding unit does not supply the load. Regarding battery charging states, three states are defined; 1-charging, -1- discharging, 0-open circuit.

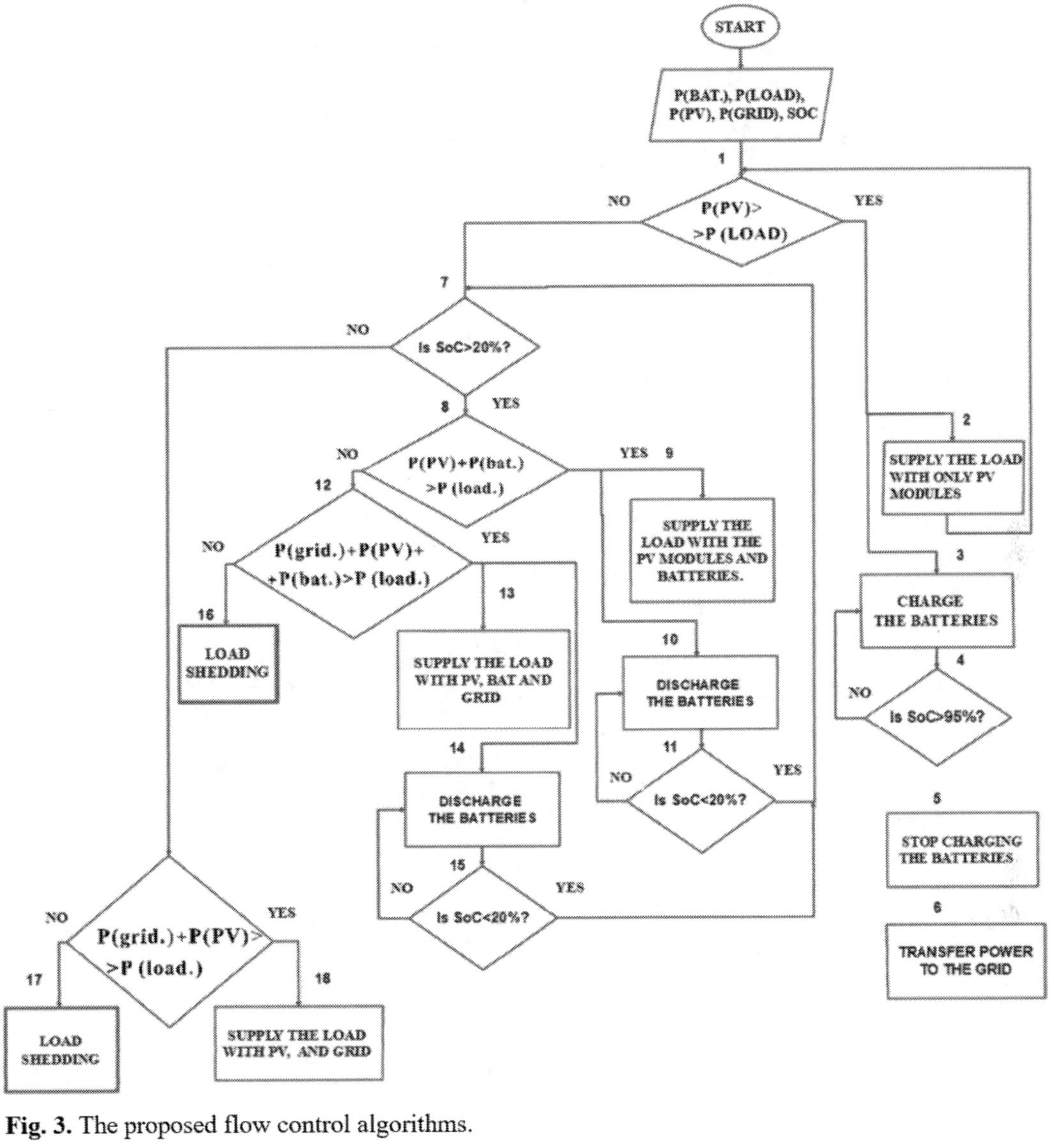

Fig. 3. The proposed flow control algorithms.

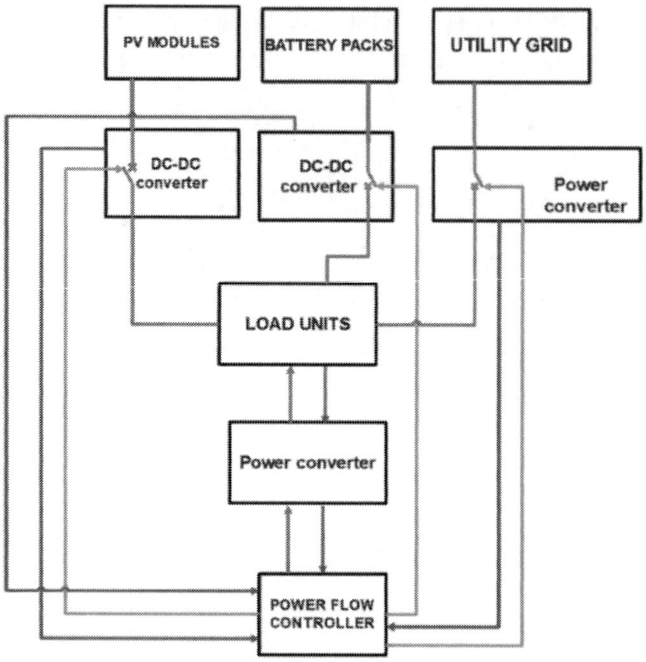

Fig. 4. Power flow control structure.

The proposed flow control employs the following procedures:

First, total power from PV units $\sum_0^N P_{PV}$, total power from battery units $\sum_0^N P_{BAT.}$, available grid power P_{GRID}, total load power $\sum_0^N P_{LOAD.}$ must be identified.

If $\sum_0^N P_{PV} > \sum_0^N P_{LOAD.}$, then PV modules are enough to supply the load with enough power without the need for battery and grid power. It is illustrated in Figure 5 where $\sum_0^N P_{PV}$ is 60W, while $\sum_0^N P_{LOAD.}$ is 40 W. The operation numbers refer to the proposed algorithm in Figure 3. In this case, battery units start charging until their SoC level is 95% to prevent overcharging. If the SoC level goes beyond the pre-defined threshold, battery units stop charging and the excessive energy is transferred to the grid. Exported energy to the grid is the following (12):

$$P_{GRID.} = \sum_0^N P_{PV.n} - \sum_0^N P_{LOAD.n}. \tag{12}$$

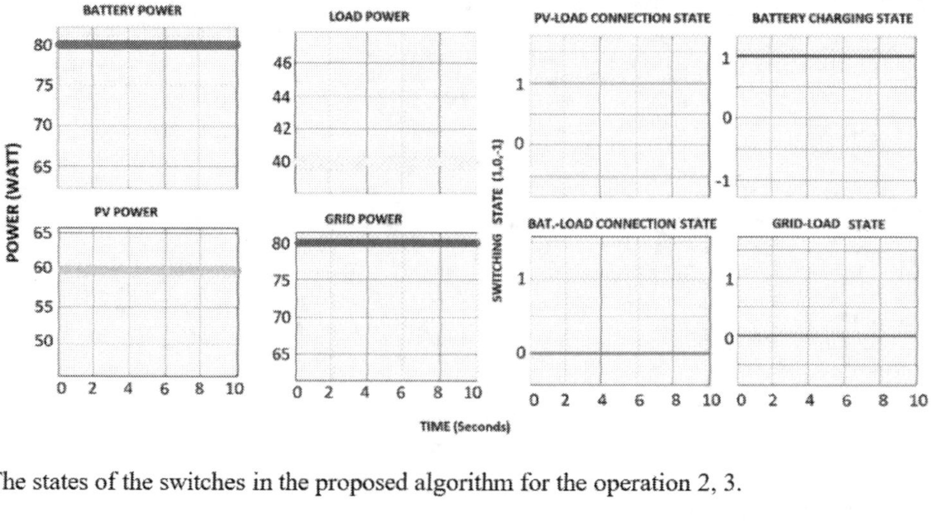

Fig. 5. The states of the switches in the proposed algorithm for the operation 2, 3.

If $\sum_0^N P_{PV}$ <$\sum_0^N P_{LOAD.}$, then PV module output power cannot match the demand power, which necessitates the additional battery connection $\sum_0^N P_{BAT.}$. It checks the SoC level of the battery whether it is higher than 20%. If this condition is true, then the condition $\sum_0^N P_{PV} + \sum_0^N P_{bat.} > \sum_0^N P_{load}$ is checked. If it is also satisfied then batteries are connected to the load and discharging process starts (state -1) until the SoC drops to below 20% . It is illustrated in the Figure 6 where $\sum_0^N P_{PV}$ is 30W, $\sum_0^N P_{bat.}$ is 80W , SoC=75%, while $\sum_0^N P_{LOAD.}$ is 70W.

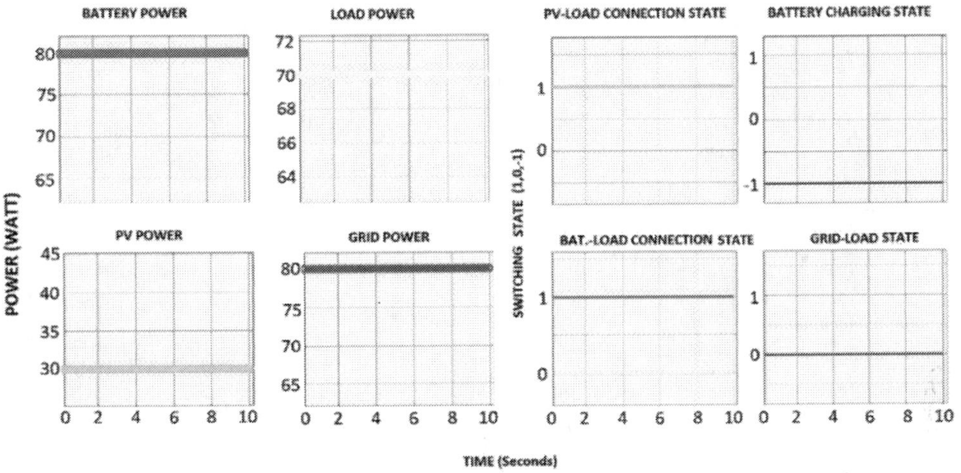

Fig. 6. The states of the switches in the proposed algorithm for the operation 9, 10.

If the battery SoC>20%, while $\sum_0^N P_{PV}$+$\sum_0^N P_{BAT.}$ cannot deliver the demand power to load, then the utility grid is switched on to be connected to the load. PV modules, batteries and the grid is connected to the load. Batteries discharge in this mode. It is illustrated in the Figure 7 where $\sum_0^N P_{PV}$ is 30W, $\sum_0^N P_{bat.}$ is 30W, SoC=75%, available $P_{GRID.}$ = 80W ,while $\sum_0^N P_{LOAD.}$ is 70W. Imported energy from the grid is the following (13):

$$P_{GRID.} = \sum_0^N P_{LOAD.n} - \sum_0^N P_{PV.n} - \sum_0^N P_{bat.n}. \quad (13)$$

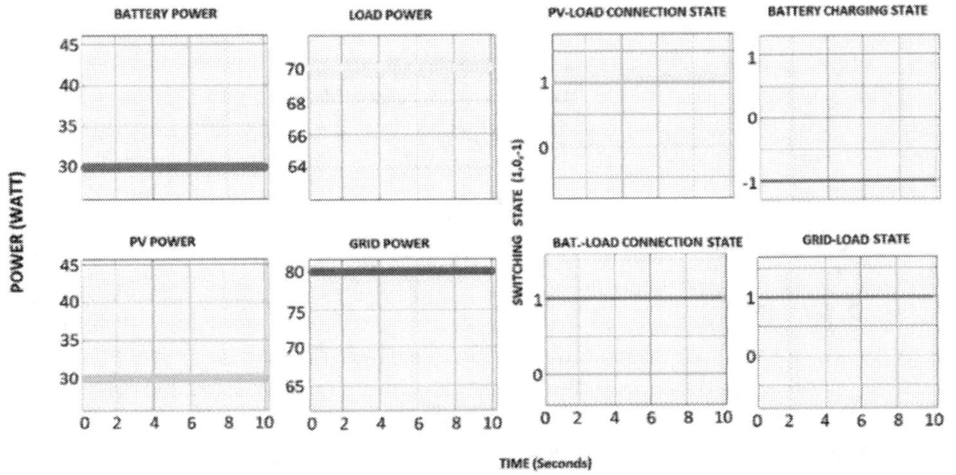

Fig. 7. The states of the switches in the proposed algorithm for the operation 13, 14.

If the battery SoC<20%, then the utility grid is switched on to be connected to the load to contribute to the PV. Batteries are disconnected in this mode because of the low SoC level.

 https://doi.org/10.1051/e3sconf/202340403001

It is illustrated in the Figure 8 where $\sum_0^N P_{PV}$ is 30W, $\sum_0^N P_{bat.}$ is 30W, available $P_{GRID.} =$ 80W ,while $\sum_0^N P_{LOAD.}$ is 70W.

Imported energy from the grid is the following (14):

$$P_{GRID.} = \sum_0^N P_{LOAD.n} - \sum_0^N P_{PV.n}. \quad (14)$$

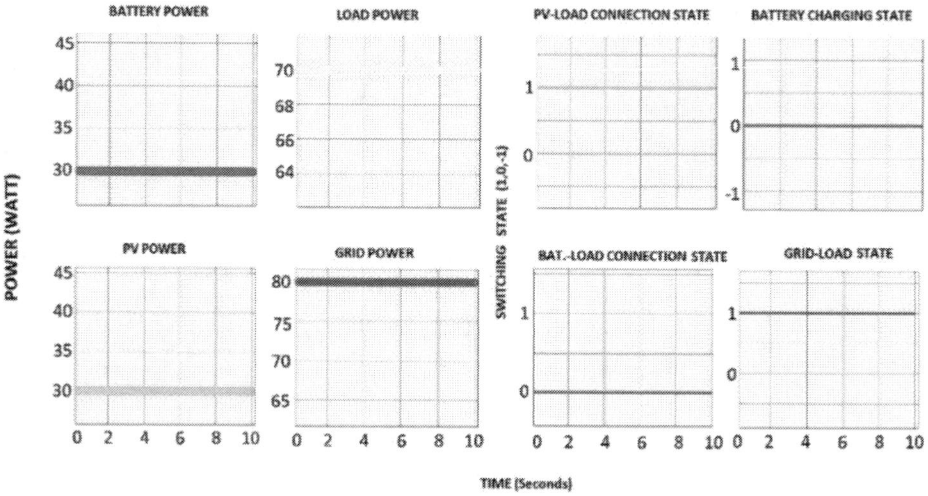

Fig. 8. The states of the switches in the proposed algorithm for the operation 18.

It is possible that the grid power is not available or not enough. Under these circumstances all power sources are switched on (unless battery SoC<20%), then the system goes to the load shedding mode, in which low priority load units are disconnected until the total power of sources can meet the demand, at least, power of high priority loads.

4 Conclusion

In conclusion, this paper highlights the importance of power flow control systems in DC microgrids with photovoltaic inputs to ensure a continuous power supply to the load. An improved power flow control method of energy sources to the load unit based on a State-of-Charge using Coulomb counting method is proposed to ensure an uninterrupted power supply. The bidirectional SEPIC-ZETA converter is chosen as an interface between the battery units and DC bus, as it allows charging and discharging of the battery units without the need for changing DC bus voltage. The proposed approach ensures that the load consistently receives the required power, and grid power is used only when the estimated combined power from the photovoltaic units and batteries is under the load demand power. A working algorithm is designed and simulations are carried out. The simulation results of the proposed approach show that it provides satisfactory control performance to meet the load demand power and can be implemented in various DC microgrid applications to enhance their performance. Overall, the proposed approach is effective in meeting load demand power and can significantly improve the performance of DC microgrid systems.

References

1. R. Kandari, N.A Mittal, Chapter 4 - DC microgrid, Microgrids, Academic Press, Pages 91-139, (2022) ISBN 9780323854634

2. M. Saad Bin Arif, M. Asif Hasan, 2 - Microgrid architecture, control, and operation, In Woodhead Publishing Series in Energy, Hybrid-Renewable Energy Systems in Microgrids, Woodhead Publishing, Pages 23-37, (2018), ISBN 9780081024935
3. S. Pannala, N. Patari, A. K. Srivastava and N. P. Padhy, "Effective Control and Management Scheme for Isolated and Grid Connected DC Microgrid," in IEEE Transactions on Industry Applications, vol. 56, no. 6, pp. 6767-6780, Nov.-Dec. (2020)
4. Z. Wang et al., "Research on the active power coordination control system for wind/photovoltaic/energy storage," IEEE Conference on Energy Internet and Energy System Integration (EI2), Beijing, China, pp. 1-5, (2017)
5. Z. Shi, W. Wang, Y. Huang, P. Li and L. Dong, "Simultaneous optimization of renewable energy and energy storage capacity with the hierarchical control," in CSEE Journal of Power and Energy Systems, vol. 8, no. 1, pp. 95-104, Jan. (2022)
6. F. O. Ramos, A. L. Pinheiro, R. N. Lima, M. M. B. Neto, W. A. S. Junior and L. G. S. Bezerra, "A Real Case Analysis of a Battery Energy Storage System for Energy Time Shift, Demand Management, and Reactive Control," IEEE PES Innovative Smart Grid Technologies Conference - Latin America (ISGT Latin America), Lima, Peru, pp. 1-5, (2021)
7. Y. R. Rodrigues, M. R. Monteiro, A. C. Zambroni de Souza, P. F. Riberiro, L. Wang and W. Eberle, "Adaptative Secondary Control for Energy Storage in Island Microgrids," IEEE Power & Energy Society General Meeting (PESGM), Portland, OR, USA, pp. 1-5, (2018)
8. J. Stuchlý and J. Vramba, "Analyses of power quality and voltage variation of photovoltaic power plant connected into 22 kV public grid," 13th International Conference on Environment and Electrical Engineering (EEEIC), Wroclaw, Poland, , pp. 48-51, (2013)
9. Z. Ben Mahmoud, M. Hamouda and A. Khedher, "Effects of series and shunt resistances on the performance of PV panel under temperature variations," International Conference on Electrical Sciences and Technologies in Maghreb (CISTEM), Marrakech & Bengrir, Morocco, pp. 1-7, (2016) doi: 10.1109/CISTEM.2016.8066772
10. B. S. Bhangu et al., "Nonlinear observers for predicting state-of-charge and state-of-health of lead-acid batteries for hybrid-electric vehicles," IEEE transactions on vehicular technology, vol. 54, no. 3, pp. 783-794, (2005)
11. M. R. A. Ratul, "Energy Management of a Renewable Hybrid Isolated DC Microgrid," Int. J. Res. Appl. Sci. Eng. Technol., vol. 9, no. 11, pp. 116–123, Nov. (2021)
12. Z. Miao, L. Xu, V. R. Disfani, and L. Fan, "An SOC-Based Battery Management System for Microgrids," IEEE Trans. Smart Grid, vol. 5, no. 2, pp. 966–973, Mar. (2014)
13. Ma, T.T.H.; Yahoui, H.; Vu, H.G.; Siauve, N.; Morel, H. A Control Strategy of DC Building Microgrid Connected to the Neighborhood and AC Power Network. Buildings, 7, 42 (2017)
14. F. Li, Z. Lin, Z. Qian and J. Wu, "Active DC bus signaling control method for coordinating multiple energy storage devices in DC microgrid," 2017 IEEE Second International Conference on DC Microgrids (ICDCM), Nuremburg, Germany, pp. 221-226 (2017)
15. E. Yusubov and L. Bekirova, "Adaptive Metaheuristic Moth-Flame Optimized Droop Control Method for DC Microgrids," International Conference Automatics and Informatics (ICAI), Varna, Bulgaria, pp. 204-209 (2022)

Pollution of Varna Lake and possibilities of using the pollutants as resources

Todor Ganchev[1], *Valentina* Markova[1], *Ivalena* Valcheva-Georgieva[1*], *and Ivan Dobrev*[2]

[1]Technical University - Varna, 1 Studentska Str., BG 9010, Varna, Bulgaria
[2]Hydro-Map EOOD, 166A Slivnitsa Blvd., 3rd floor, office №17, Varna, Bulgaria

Abstract. Varna Lake is of great social and economic importance for the city of Varna and the entire region. The urbanisation of the areas surrounding the Varna-Beloslav Lake complex and the accelerated industrialisation around the coastline of Varna Bay have significant anthropogenic pressure on the aquatic ecosystem. An essential aspect is that industrial development, maritime transport and urbanisation are substantial sources of large-scale inorganic and organic pollution with all the resulting negative consequences. Recent measurements show that the surface sediments, which are a source of nutrients for the water body of Varna Lake, are contaminated with heavy metals and oil products. In this paper, we present a quantitative assessment of the degree of contamination with heavy metals (Hg, As, Fe, Cu, Pb) and petroleum products in the upper sediments of Varna Lake and then discuss opportunities for using some of these pollutants as a resource. The exploitation of these resources would bring significant benefits for improving the ecological status of Varna Lake.

1 Introduction

Varna Lake is located within the borders of the northern Bulgarian Black Sea coast. Until the digging of the first canal between the Varna and Beloslav Lakes in 1923, these lakes were a single lake basin. According to its genesis, the lake belongs to the seaside lakes [1]. Two artificially excavated navigation channels connect the lake to Varna Bay. Lambev et al. carried out measurements in Varna Lake and determined an area of 15.7 km2, a maximum depth of 21.3 m, a maximum length of 10.5 km and a maximum width of 2.5 km [2].

The Varna-Beloslav Lake complex has significant social and economic importance for the surrounding areas. At the same time, the anthropogenic influence on the lakes is increasing with the acceleration of urbanisation and industrialisation of Varna Bay coastal areas. Specifically, industrial development, maritime transport and urbanisation are identified as the primary sources of heavy inorganic and organic pollution. This leads to the deterioration of the ecological condition of the water body of the Varna-Beloslav Lake complex, which refers to the eutrophic type of lakes. During the circulation of substances in these lakes, part of the produced organic substances is preserved without being mineralised,

* Corresponding author: ivalena.georgieva@tu-varna.bg

accumulating mainly in the bottom sediments. Besides organic substances, heavy metals (mercury (Hg), arsenic (As), iron (Fe), copper (Cu), lead (Pb)) and petroleum products accumulate in the sediments.

In this report, we present a quantitative evaluation of the degree of contamination of the sediments of Varna Lake with the above-mentioned heavy metals and petroleum products. Conceptual ideas for cleaning the lake sediments are discussed, as well as methods for cleaning which consider these pollutants as a resource. Such a cleaning, in turn, will support the sustainable management of water resources in the area.

2 Methods

From October 2021 to February 2022, 140 bottom samples were taken from the surface sediments of Varna Lake. Each sample was taken 40 cm below the bottom level. Sampling was carried out by qualified diving teams in two pre-selected target areas (Zone 1 and Zone 2), according to a predetermined measurement grid with a distance between sampling sites of 200 m [3]. Zone 1 covers the eastern parts of Varna Lake and the two navigation channels connecting it to Varna Bay. Zone 2 is located in the central part of Varna Lake, near the village of Kazashko Figure 1.

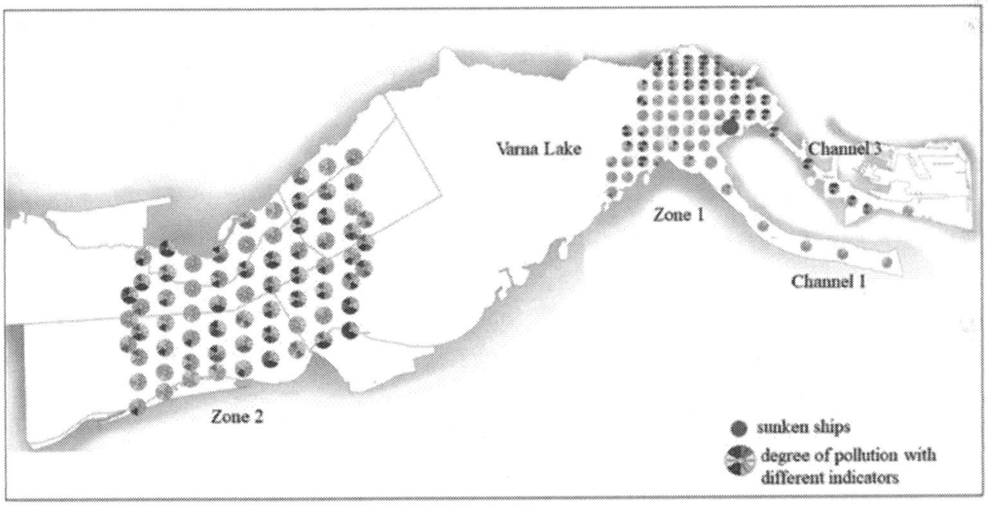

Fig. 1. Extent of the researched area, according to the Hydro Map with additions.

The bottom sediment samples were taken at a maximum working depth of 19.5 m below the water surface, with visibility of 0 ÷ 1.40 m and a water temperature of 12 °C. All 140 samples were subjected to chemical analysis to assess contamination levels with heavy metals and petroleum products. The chemical analysis was carried out by Eco-Consult-Engineering OOD - an accredited laboratory in Burgas. This analysis aims to evaluate metal tracers such as mercury (Hg), arsenic (As), iron (Fe), copper (Cu), lead (Pb) and petroleum products. Identifying the metal indicators mentioned above: Iron/Fe, Lead/Pb, Arsenic/As, Copper/Cu, and Mercury/Hg was made according to the procedures described in the Bulgarian national standard -- test method BDS EN 16170:2016 (https://bds-bg.org/bg/project/show/bds:proj:99534). The identification of petroleum products in the probes was carried out according to the procedures defined in the Bulgarian national standard BDS EN ISO 16703:2011 (https://bds-bg.org/bg/project/show/iso:proj:39937). In these analyses, we considered the commonly accepted unit of measurement for the indicators -- mg.kg^{-1}.

The reference values used in the subsequent assessment of contamination levels follow the International Atomic Energy Agency (Reference sheet IAEA-158/31.08.2008 and Reference sheet IAEA – 459/10.06.2017) (Reference sheet IAEA-158/31.08.2008 and Reference sheet IAEA – 459/10.06.2017). The reference sheets have no aggregate reference values for petroleum products but are broken down by type. Therefore, when preparing the legend for petroleum products, the reference value was taken based on the average value of the laboratory samples, and the rest were defined as contaminated and uncontaminated areas [3].

3 Results

Varna Lake is defined as highly modified. As a result of human activity, its physical characteristics have undergone significant change. The quality of the aquatic environment is determined by the ecological potential of the lake itself. We can consider three categories of ecological potential - good, medium, low and bad, depending on the level of effort invested in environment protection. The environmental assessment of the waters of Varna Lake is within the limits of low potential, with minor exceptions near the lake-bay contact zone, where it is medium. In recent years, due to a drop in production in the chemical complex in Devnya, the TPP (Thermal power plant) Varna and the Beloslav glass production factory, there has been a decline in the influence of industrial activity in the area. The maintenance of the state of the lake system in "low-poor quality" of ecological potential is due to the loading and unloading activities of the ports, the waters from the Provadiyska River, the reduced capacity of the treatment plants, the increasing urban population, etc. In general, the ecological potential of the lake varies from good to bad, with 52% having an average, 27% good and 19% bad potential [4]. The polluted sediments and the dredged material depot located in the water area of the lake have a significant impact on its ecological condition.

The surface sediments are an essential source of nutrients for the water body of Varna Lake. The nutrients released from the sediments exceed the total external load and are sufficient to support hyper eutrophication. Even with significant reductions in point source inflow, benthic streams will continue to support high primary production and subsequent organic loading. This, in turn, will cause nutrient recycling leading to enhanced primary production. This suggests there needs to be more opportunity for the Varna-Beloslav Lake complex to meet WFD requirements for good condition [5].

At present, the hydrochemical characteristics of Varna Lake are determined by its connection with Beloslav Lake and Varna Bay. This is because Beloslav Lake receives polluted industrial and domestic wastewater, through which heavy metals, petroleum products and detergents enter the lake. The lead, copper, arsenic and cadmium concentrations are higher than in the bay [6].

After an analysis of the average contamination of all samples from the two areas of interest, it was observed that the most significant accumulation of iron, lead and copper is in Zone 1, and the highest pollution in Zone 2 is due to arsenic, mercury and petroleum products Figure 2. The accumulation of these heavy metals and petroleum products in the sediments of Varna Lake is due to the solid industrial pressure, urbanisation and shipping [3]. After a comparison of the average values of pollution in the two zones, it was observed that the indicators - mercury, copper and petroleum products prevail over the others.

 https://doi.org/10.1051/e3sconf/202340403002

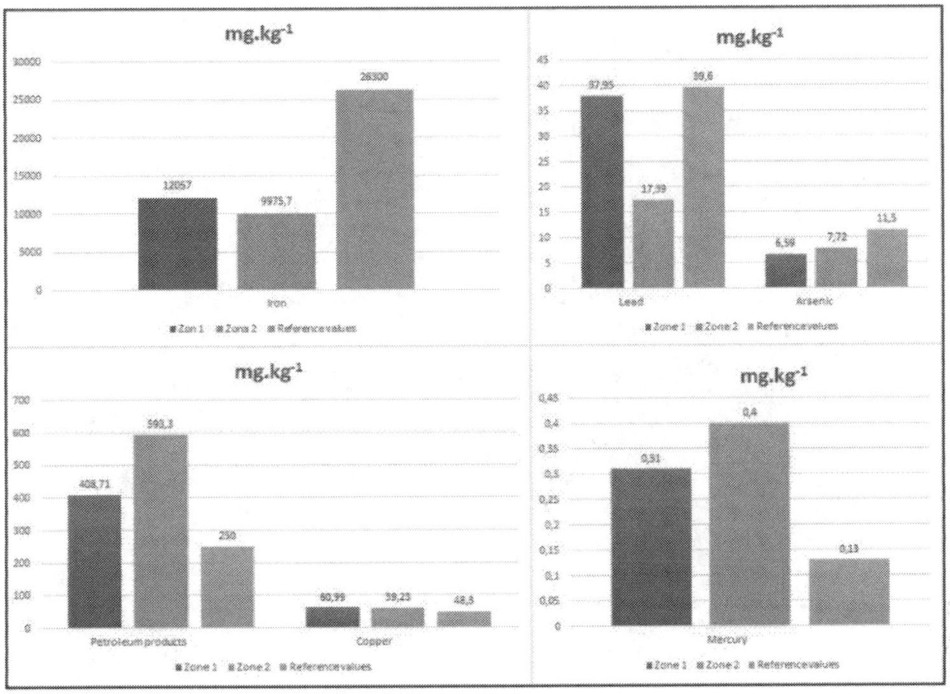

Fig. 2. Average pollution values in Zone 1 and Zone 2.

Finer-textured sediments with higher organic content accumulate higher concentrations of heavy metals than coarse sediments. This depends on the specific hydrodynamic conditions influencing the distribution of fine particles, such as silt and clay. A large part of the heavy metals are retained in the slag of the treatment plants but also easily reach drinking water [7]. In marine areas, such as Varna Lake, which have a low deposition energy, the accumulation of fine particles and pollutants is facilitated, while in those characterised by a high deposition energy (waves, currents), coarse particles (sand) prevail. The speed of the currents in Varna Lake is relatively low, and fine sediments are deposited, in which the pollutants are retained and accumulated.

It has been established that there is a constant intensive water exchange between Varna Lake and Varna Bay, where seawater mainly intrudes through Channel 1 in the west direction, and lake waters are directed to the bay through the old channel. This direction of the lake-bay water exchange may account for the more significant pollution in the Old Channel, as more heavily polluted lake waters pass through it in the direction of the bay. Another reason for the difference in the contamination could be that in Channel 1, periodic dredging of the bottom is carried out to provide a navigable depth Figures 3 and 4. Given the direction of movement of the lake waters and the location of the dredged material dump, there is a possibility that fine sediments and their pollutants are transported to the Old Channel [3].

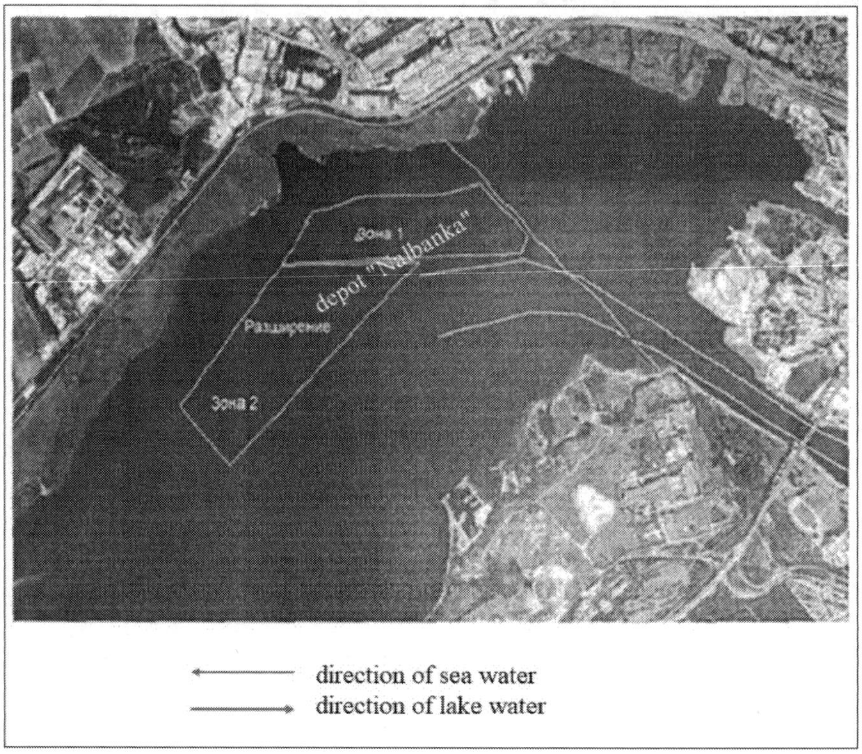

Fig. 3. Location of a landfill for dredged sediments in Varna Lake [8].

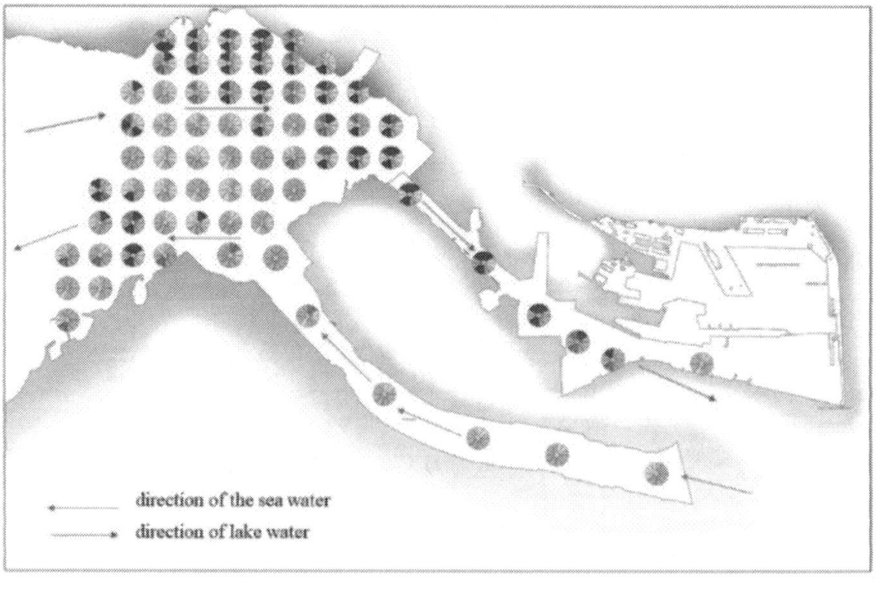

Fig. 4. Pollution in Zone 1 and direction of water exchange (lake-bay), according to the Hydro Map, with additions.

All accumulated pollutants in the sediments of Varna Lake are continuously released into the water body and contribute to maintaining its poor ecological condition. This, in turn, directly impacts the normal functioning of the ecosystem and biodiversity and is a prerequisite for the accumulation of pollutants in the body of aquatic inhabitants and the subsequent impact on human health. Restoring the self-purification function of the water bodies will improve their ecological status, contributing to their sustainable management.

In addition to pollution with heavy metals, petroleum products and eutrophication, there are remains of sunken ships in the Varna Lake area, which, in addition to pollutants, can hinder navigation Figure 5. It is necessary to survey and document all ship remains and consider measures for their removal and subsequent recycling. The use of ship remains for raw material and recycled iron will help clean the lake's water body.

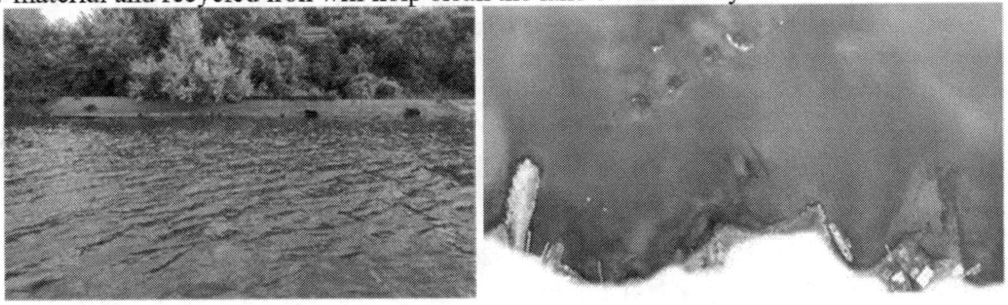

Fig. 5. Remains of sunken ships in the Varna Lake area. (a) Iron fragments visible above the water surface. (b) Sinken fragments below the surface.

In Varna Lake, near the Kazashko village, the presence of estuarine healing mud has been established. It has been formed for thousands of years and is used for treatment in balneology. After 1978, Vladeva et al. conducted annual surveys of the deposit and found changes in the pH of the healing mud and the presence of mercury in some areas [9]. By bringing the lake to a good ecological condition and protecting the site from pollution, there will be favourable conditions for developing balneological tourism. This, in turn, will lead to new jobs being created for the population.

Various methods can be used to prevent the negative consequences on Varna Lake and bring it into a good ecological state, which will favour the blue and circular economy by turning waste (pollutants) into a resource.

Different methods are used for phytoremediation and bioremediation of sediments contaminated with heavy metals, petroleum products and organic substances. They use plant species, zeolites, microorganisms, etc., to remove contaminants. The removal of the dredged sediment dump will have a beneficial effect on the ecological status of the lake. The sediments can be subjected to phyto- and bioremediation and used as a bulk material to raise the level of lake shores Figure 6. The improvement of the ecological condition of the water body of Varna Lake will have a favourable impact on the development of various forms of alternative tourism (fishing, educational, spa, etc.).

Fig. 6. Bulk works along the shore of Varna Lake [10]

Biomass can be used for heating, electricity production and transport of biofuels. It is a natural, completely self-renewable and essential alternative source of energy. Biomass is obtained from different types of organic matter: energy plants, forests, agricultural and urban waste, including wood and household waste, and others. In Bulgaria, there is a great potential for biomass suitable for biogas production [11, 12].

The common reed (*Phragmites australis*) is a characteristic herbaceous plant along the shores of the entire lake complex Figure 7. Its presence contributes to water purification and sediments on the coasts. On the one hand, it can be used for phytoremediation by expanding its areas; on the other hand, the biomass (plant waste) can be used to produce biofuels and energy. Using biomass for energy purposes practically does not pollute the environment, and therefore it is considered a source of ecologically clean or "green" energy. In October 2022, one of the chemical industry enterprises in the Solvay Sodi region developed and presented a project to construct an installation for the thermal treatment of alternative fuels from non-hazardous waste and biomass. In this way, the enterprise will ensure long-term competitiveness and sustainability, improve regional air quality, help in waste management, etc. After building the plant, the company could use the biomass from the reed areas.

Fig. 7. Phragmites australis

Fig. 8. Sewage drainage channel of the Padina tailings pond

Wastewater from the chemical industry complicates the ecological condition of the water body. It is necessary for enterprises to be stimulated and supported in investments related to the use of recycled wastewater in the production process. This will reduce pollutants introduced into the lake and water costs, as businesses will repeatedly recycle and reuse the processed water in production Figure 8.

Fig. 9. State of the water body

The presence of elements such as nitrogen and phosphorus in large quantities causes eutrophication of the water body, destroying aquatic organisms Figure 9. Phosphorus is an exhaustible natural resource. Its presence in the water body and sediments should be reassessed because its extraction can provide a recycling solution. This calls for cross-sectoral cooperation and cost-effective ways to recover and reuse phosphorus and other resources from waters and sediments [13].

It has been proven that the cultivation and use of algae as a source of biofuels has numerous benefits for the environment and enables the development of a circular and blue economy. Algae is a versatile feedstock that can be used to refine various end products such as biodiesel, bioethanol, biomethane and biohydrogen. There are opportunities to build a waste-free technology for the simultaneous production of electricity and biofuels. The produced biofuel does not contain sulphur and is non-toxic, biodegradable and relatively harmless to the environment. Algae cultivation requires land unsuitable for agriculture, carbon dioxide, solar radiation, and polluted water, which the algae use as food. The nitrates, ammonia and phosphates that pollute the water serve as nutrients for the algae. In addition, algae are great biofixers – one metric ton of biomass absorbs 1.6 t of CO_2, producing 1.2 t of oxygen [14]. They take CO_2 from the atmosphere to use as a form of energy for themselves. Linking algae cultivation for biodiesel with the production of wastewater and the release of CO_2 is a perfect opportunity to protect lake waters, improve air quality in the area and create new jobs.

Important for the Bulgarian maritime economy are sectors such as maritime tourism, aquaculture, maritime transport, port activities, shipbuilding, ship repair, etc. These sectors are the basis of the "Blue Economy" of the Black Sea regions of Bulgaria and depend on the quality of natural ecosystems and coastal and marine resources. That is why it is essential to maintain a good ecological condition of Lake Varna for the normal functioning and development of the circular and blue economy and the sustainable management of water resources.

4 Conclusions

As a result of the conducted research on sediments in Varna Lake, elevated values of mercury (Hg), arsenic (As), iron (Fe), copper (Cu), lead (Pb) and petroleum products were found. Their accumulation in the sediments of Varna Lake is due to the solid industrial pressure, urbanisation, shipping, polluted waters from inflowing rivers, reduced capacity of treatment plants, etc. When comparing the average pollution values in the two studied areas, it was found that mercury, copper and petroleum products prevail over the others. The ecological assessment of the waters of Varna Lake is within the limits of low potential, with minor exceptions near the lake-bay contact zone, where it is medium. Of great importance for their sustainable management is their introduction into a good ecological condition and restoration of self-purifying functions.

Phytoremediation, bioremediation of sediments contaminated with various heavy metals, petroleum products and organic substances can be mentioned as good practices for cleaning the sediments and the water body; wastewater recycling; cultivation and use of algae as a source of biofuels; extraction of phosphorus from polluted areas and its reuse, etc. Various methods can be used to prevent the negative consequences on Varna Lake and bring it into a good ecological state, which will favour the blue and circular economy and sustainable management of water resources by turning waste (pollutants) into a resource.

This research was carried out with the financial support of the KnoWat project, entitled, "Enhancing knowledge about the state of the aquatic environment in Lake Varna," funded by the Maritime and

Fisheries Programme of the Ministry of Agriculture of Republic of Bulgaria with grant №BG14MFOP001-6.004-0003.

References

1. T. Krastev, St. Stankova, Natural geography of Bulgaria and the Black Sea. Part I - General characteristics of the national geographical space. Sh., UE"E. Konstantin Preslavsky", ISBN 978-954-577-414-0, p. 240, (2008)
2. T. Lambev, B. Prodanov, L. Dimitrov, I. Kotsev, *Digital terrain model of the Varna and Beloslav Lakes, North Bulgarian Black Sea Coast*, Proc. SPIE 11524, Eighth International Conference on Remote Sensing and Geoinformation of the Environment (RSCy2020), 1152422 (26 August 2020), doi: 10.1117/12.2571104, (2020)
3. T. Ganchev, V. Markova, I. Valcheva-Georgieva1, I. Dobrev. *Assessment of pollution with heavy metals and petroleum products in the sediments of Varna Lake*. REVIEW OF THE BULGARIAN GEOLOGICAL SOCIETY, vol. 83, part 2, 2022, ISSN 0007-3938, pp. 3–9. https://doi.org/10.52215/rev.bgs.2022.83.2.3, (2022)
4. K. Stefanova, *Diagnostics of the ecological potential in the Beloslav-Varna lake system and assessment of the environment according to zooplankton indicators*, Journal of the Union of Scientists - Varna, Section "Marine Sciences", p. 28, (2014)
5. V. Doncheva, *Nutrients in pore water from surface sediment layer along the eutrophication gradient (Varna Lake – Varna bay case study)*. Comptes rendus de l'Acad´emie bulgare des Sciences, Tome 63, No 4, pp. 547-554, (2010)
6. G. Shtereva, *Hydrochemical characteristics of Varna Bay and Varna Lake*, In: 10 books about Varna, Book 5, Ed. Sea World, Varna, pp. 320-326 (2006)
7. L. Lazar, et al., Black Sea state of environment based on ANEMONE Joint Cruise. CD Press, ISBN 978-606-528-530-9, p. 187. http://www.blacksea-commission.org/Downloads/ANEMONE/ Deliverable% 202.3.pdf, (2021)
8. https://riosv-varna.org/Prevantivna%20dejnost/os/2021/1-DOS.pdf
9. http://banispa.com/2011/03/%D0%BB%D0%B5%D1%87%D0%B5%D0%B1%D0%BD%D0%B0-%D0%BA%D0%B0%D0%BB-%D0%BE%D1%82-%D0%B2%D0%B0%D1%80%D0%BD%D0%B5%D0%BD%D1%81%D0%BA%D0%BE%D1%82%D0%BE-%D0%B5%D0%B 7%D0%B5%D1%80%D0%BE/
10. https://ecovarna.info/8751-2/
11. A. Bankova, *Possibilities and problems in using biomass as an energy source*, IIII International Scientific Congress. 50 years of Technical University - Varna, ISBN 978-954-20-0550-6, pp. 105-109, (2012)
12. A. Bankova, *Advantages of waste utilization for the production of biofuels (biogas)*, XVII Scientific - technical conference with international participation "Eco Varna 2012", Vol. -19, ISBN-954-20-00030, pp. 27-34, (2012)
13. https://vesistosaatio.fi/jarvien-kunnostuksessa-kohti-ravinteiden-poistoa-niiden-pidattamisen-sijaan/?fbclid=IwAR1FOvBar7c2gikJZ8bFaX2WYlD8eE1Huo9DCU4q_liCFOa1VPWdpIV8BJU
14. S. Asparuhov, Ecological, technological and structural advantages of enterprises for the production of biofuels from algae. https://www.researchgate.net/publication/333661859, (2012)

Evaluation of waste in seismic metamaterial applications

Volkan Akdogan [1*], *Selcuk* Kacin[2], *Umur* Sevim[2], *Muharrem* Karaaslan[1], and *Mustafa* Başar[3,4]

[1]Iskenderun Technical University, Dept. of Electrical and Electronics Eng., Hatay, 31200, Turkey
[2]Iskenderun Technical University, Dept. of Civil Eng., Hatay, 31200, Turkey, 31200, Turkey
[3]Amasya University, Vocational School of Tasova Yüksel Akin, Department of Motor Vehicles and Transportation Technologies, Amasya, 05800, Turkey
[4]Iskenderun Technical University, Dept. of Mechanical Eng., Hatay, 31200, Turkey

Abstract. Within the scope of this study, a simulation study was carried out in order to prove the usability of waste in seismic metamaterial studies. In the study, a square array field application was preferred, and a 3-layer cylindrical pile design was used. In addition, direct contact of waste with soil and direct air is prevented. Within the scope of the study, polypropylene, which is frequently contained in medical products, concrete as a containment layer, and lime materials to prevent leakage of hazardous waste were used as materials. In addition, a design has been made within the soil structure as the ground structure. As a result of the study, it was determined that transmission losses occur in low frequency regions such as 3-10 Hz values due to obtaining partial band gaps. In addition, when looking at the propagation of the vibration waves in the field plane depending on the time, it is seen that the waves are significantly reduced, and the results are promising.

1 Introduction

Earthquakes are one of the most dangerous natural disasters that have caused many losses of life and property from past to present. Millions of earthquakes occur every year. However, it is known that more than half of the natural disasters experienced are earthquakes [1-3]. Earthquake waves can be described as energy flows that cause rapid agitation on the surface and cause permanent damage to structure by generally spreading in all directions [4]. Seismic metamaterial designs have recently gained significant momentum to reduce the effects of unavoidable earthquakes [5]. Metamaterials are known as materials with superior physical properties, specially designed for engineering applications compared to normal materials, due to their structure [6-8]. First of all, in order to provide seismic insulation, columns, foundations etc. within the structure. The durability of the structure is of great importance in the elements. However, studies show that this can be achieved by using seismic metamaterials regardless of the structure to be protected [9]. Considering the studies carried out to provide seismic isolation, it is seen that optimization studies have

* Corresponding author: volkan.akdogan@iste.edu.tr

been carried out on the physical properties of the materials generally used in designs, the field sequences of the seismic metamaterials and the geometric shapes of the produced metamaterials [10-11]. In this context, when a general evaluation is made, it can be said that numerous designs can be made in seismic metamaterial designs. It is also clear that it is an area of application that needs to be developed effectively. When looking at the materials generally used in seismic metamaterial designs, it is seen that materials such as concrete, steel and rubber, which are widely used in civil engineering, are preferred [12-14].

On the other hand, this situation brings along quite high costs, considering that large pile designs should be made in order to obtain effective results in low frequency bands. Therefore, there is a need for seismic metamaterials that can work efficiently in low-frequency bands economically [15]. This situation supports that waste structures should be used in metamaterial designs as materials. Considering the pandemic conditions that have occurred in recent years in the light of these precursors, a high amount of medical waste has occurred due to Covid-19 [16]. Especially within the scope of this process, the use of masks and the rapid consumption of disposable products such as virus detection kits have triggered the increase in medical waste [17-18]. It is also known that the masks used are produced from various plastic-based polymers and various inorganic materials [19].

In this case, the idea of using waste materials in the contents of metamaterial structures arises. The main idea of this study is to ensure that the waste groups, which usually cause high carbon emissions when they are released to nature or disposed of, can be eliminated by using seismic metamaterials and that seismic vibrations can be reduced with the help of these wastes.

Within the scope of the study, buried pile designs where waste can be safely stored were made. In addition, a lime layer has been added to the design to provide extra protection. Within the scope of the study, the progress of the seismic waves over the created area with time was monitored instantly, depending on the transmission losses obtained in the frequency domain. Studies have been carried out in the frequency range of 3-10 Hz.

2 Material and method

Within the scope of this simulation study, a seismic metamaterial design by using waste has been tried to be realized. Within the scope of this study, physical properties of the plastic derivative polypropylene material as a representation of waste will be included in the simulation studies. The modulus of elasticity, density and poisson ratios of polypropylene and other materials used in the study are shown in Table 1.

Table 1. Materials used in the study and their physical properties [20-24].

Sample	Density	Elasticity module	Poisson ratio
Soil	1800	20e6	0.3
Concreate	2500	30e6	0.25
Limestone	1200	25e9	0.19
Polypropylene	905.32	4e9	0.41

Simulation studies were carried out between 3-10 Hz frequency values. In addition, the impact force of 1 N created within the scope of the study is applied to the structure by creating a 1-second distribution parameter. Time precision was created with a precision of 0.01 seconds. Depending on the frequency and time, the progress of the wave propagation was followed instantly. From the wave source, evaluations were made depending on the progress of the wave distances.

A design was made as shown in Figure 1-a and the positioning of 3-layered circular cross-section cylindrical structures in the soil was realized. The diameter of the circular section filled with the innermost polypropylene was determined as 20 cm, the diameter of the circular section defined as the lime structure immediately surrounding it was determined as 30 cm, and the diameter of the circular equation defined as the outer casing layer was determined as 40 cm. Since hazardous wastes are used here, a layer of lime has been added to the simulation studies. In addition, in order to prevent the contact of the waste structure with air starting from the ground surface, a 20 cm thick square concrete attachment was used to connect the circular structures to each other from the upper part and the contact of the waste structures with air was prevented. The top view of the circular column structures formed in the form of a square arrangement under the concrete square plate is as shown in Figure 1-b. However, looking at the characteristics of a single cell in this square sequence, the unit cell structure in Figure 1-c is obtained. In line with the data obtained here, the direction of the waves is moving as shown in Figure 1-d.

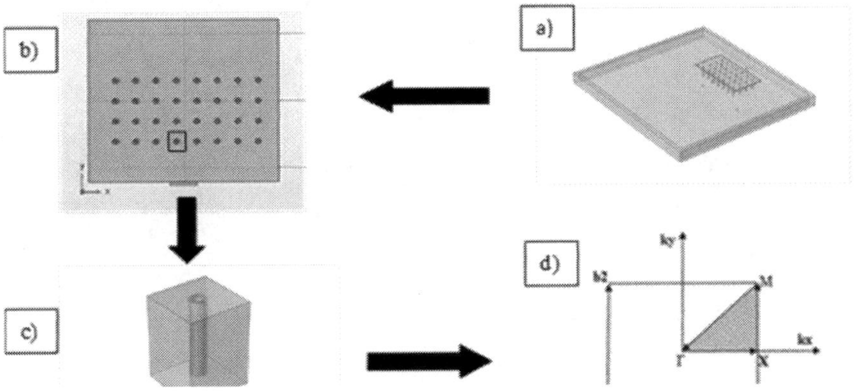

Fig. 1. Within the scope of the study, a) geometric design, b) applied square lattice sequence, c) unit cell structure and d) direction of waves on the unit cell formed.

The wave source was placed at point A at a certain distance from the design realized as shown in Figure 2. With an equal distance to this source, point B from points B and C was placed on the empty earth ground parallel to the source, and point C was placed behind the realized design. Here, the results are evaluated depending on the wave movements at the B and C points.

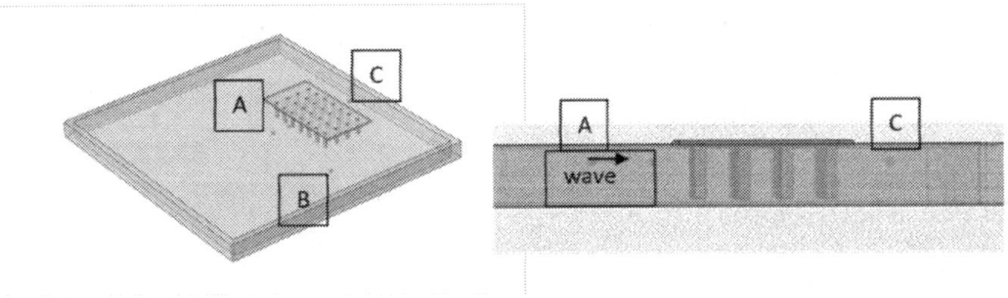

Fig. 2. Measurement points determined in simulation studies and location of wave source.

Looking at the band structure diagram shown in Figure 3, unfortunately, a complete band gap could not be obtained. However, it is seen that partial band gaps are obtained in the frequency regions of 25 Hz and higher.

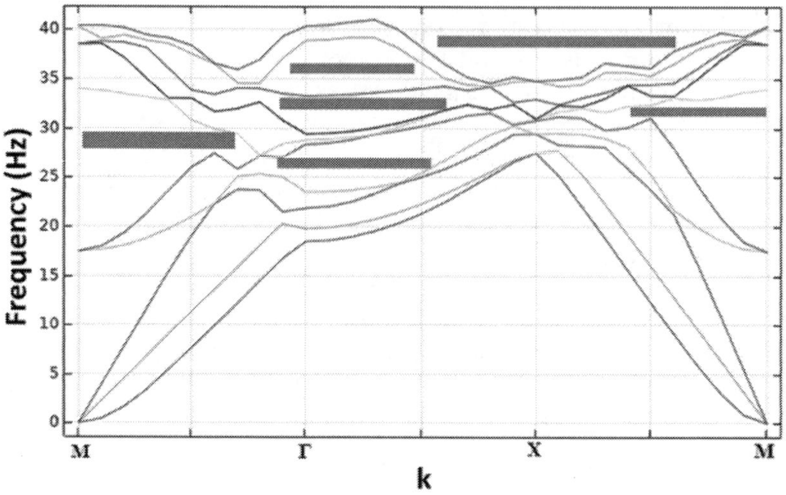

Fig. 3. Band structure obtained in simulation studies.

In addition, a striking element is that these band gaps are distributed in almost every direction. Based on this, it can be said that although it is not possible to dampen seismic effects in a full frequency range, there may be some damping in various frequency waves.

3 Results and discussions

In this section, based on the data obtained in the material and method section, the efficiency of the designed structure will be discussed by looking at the transmission losses and the progress of the waves in the time and frequency domain.

Considering the transmission losses shown in Figure 4, it is seen that in general, transmission losses occur at certain frequency points between 3Hz and 10Hz frequency value. Significant transmission losses occurring at local frequency values such as 3.5 Hz, 5.8 Hz, 8.3 Hz, 8.8 Hz, 9.3 Hz and 9.5 Hz are among the important outputs. Although there is no continuity in transmission losses in general, the restriction of transmission at point frequency values creates an opinion that the proposed structure can reduce seismic effects.

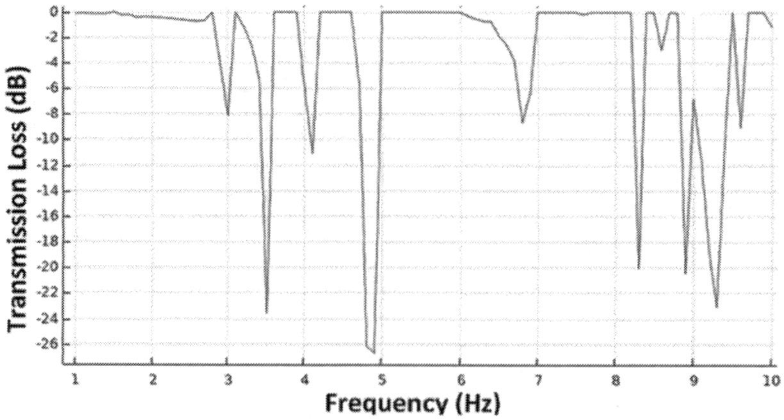

Fig. 4. Transmission losses between the proposed seismic metamaterials and the empty field.

However, the fact that the largest decibel losses occurred at 5.8 Hz is that wave propagation does not behave linearly with frequency. The region where transmission losses occur with more frequent intervals are between 8-10 Hz values. In this case, it creates the opinion that the proposed structure can function functionally in a narrow band such as 8-10 Hz, except for point values.

Looking at the results obtained in Figure 5, instantaneous emission amounts occurring at the frequency values specified in some time periods are given. Here, it is seen that the surface waves can pass behind the proposed structure in a very small amount, and seismic waves can be restricted to a significant extent in all the frequency values that occur in the transmission losses in general.

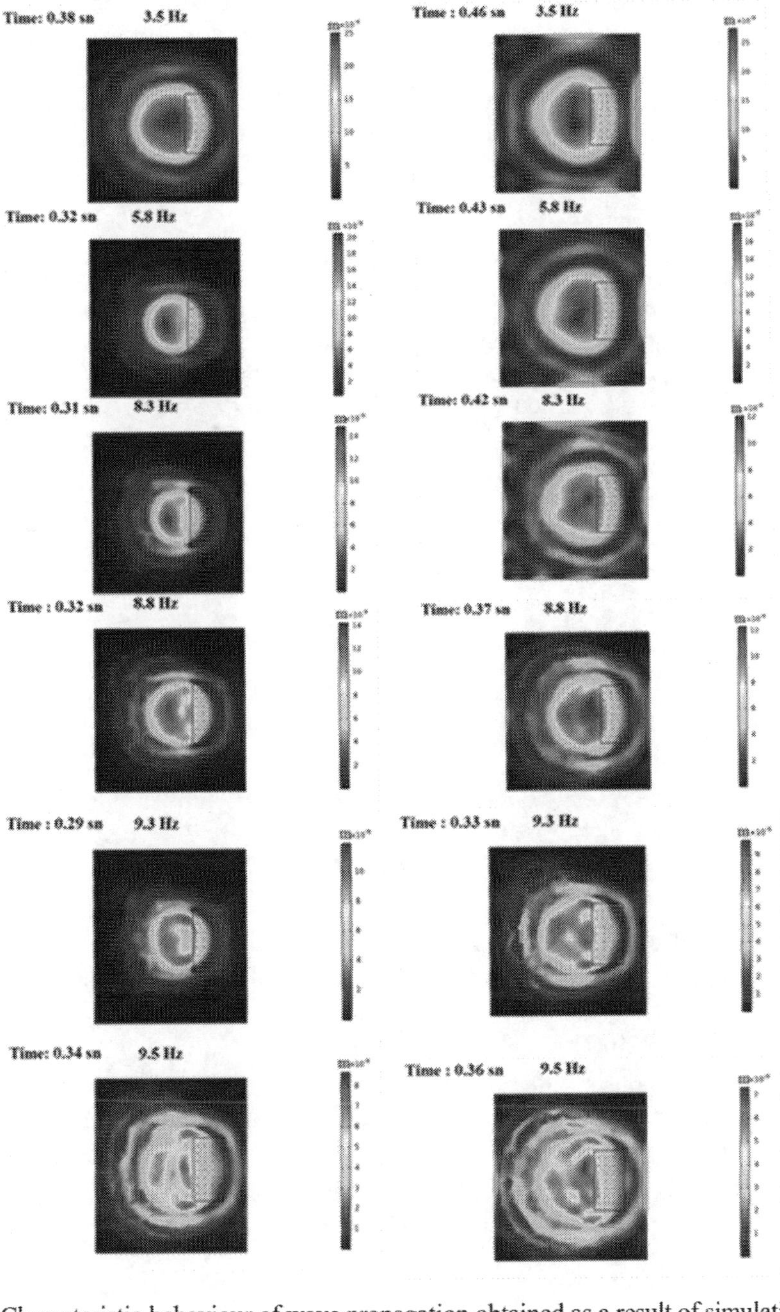

Fig. 5. Characteristic behaviour of wave propagation obtained as a result of simulation studies.

In general, as the frequency values increase, it is seen that the wave propagation increases more, however, the proposed structure, especially since the formation of the second order, cannot pass the waves. In general, it is seen that wave propagation increases significantly as time increases for all frequency values. However, remarkably, the fact that the wave motions in the proposed structure have been interrupted to a great extent regardless of the time parameter reveals that the proposed structure can be successful at these frequency values.

In addition, it is observed that as the frequency values increase, the wave propagation decreases. In this case, the results obtained especially at 9.5 Hz and 9.3 Hz values are able to absorb seismic effects at a good level. Finally, although wave propagation at 5.8 Hz was seen to be high, good seismic damping was achieved as indicated in transmission losses.

The similar characteristics of all results are generally seen as confirming each other in terms of simulation results. In addition, the fact that the wave propagation spreads linearly at each frequency value depending on the time proves that the proposed structure works reliably.

4 Conclusion

In general, the reusability of hazardous waste in metamaterial applications seems appropriate within the scope of the study. In addition, it can be a great advantage economically if waste is used as a replacement of conventional material.

In particular, millions of tons of waste are unrecoverable and trigger vital environmental problems. Instead of increasing carbon emissions by disposing of waste structures, it has been proven that these waste structures can be safely used in seismic metamaterials to reduce seismic effects.

Within the scope of the study, it has been shown that approximately 1.8 m3 of waste can be safely stored in an area of 90 m2. As a result, it is promising that both the mitigation of earthquake effects and the efficient implementation of waste recovery policies have been proven.

This research was supported by the Disaster and Emergency Management Presidency of Turkey (UDAP-Ç-19-21) for the financial support and presentation support by 2224-A Grant Program for Participation in Scientific Meetings Abroad with application number of 1919B022303039.

References

1. S. K. Duggal, Earthquake resistant design of structures, New Delhi: Oxford university press. (2007)
2. R. Bilham, Lessons from the Haiti earthquake, Nature, 463(7283), 878-879 (2010)
3. Y. Achaoui, T. Antonakakis, S. Brûlé, R. V. Craster, S. Enoch, & S. Guenneau, *Clamped seismic metamaterials: ultra-low frequency stop bands*, New Journal of Physics, 19(6), 063022 (2017)
4. Y. Achaoui, B. Ungureanu, S. Enoch, S. Brûlé, & S. Guenneau, *Seismic waves damping with arrays of inertial resonators*, Extreme Mechanics Letters, 8, 30-37 (2016)
5. R. Cai, Y. Jin, T. Rabczuk, X. Zhuang, & B. Djafari-Rouhani, *Propagation and attenuation of Rayleigh and pseudo surface waves in viscoelastic metamaterials*, Journal of Applied Physics, 129(12), 124903 (2021)

6. S. Kacin, M. Ozturk, U. K. Sevim, B. A. Mert, Z. Ozer, O. Akgol, & M. Karaaslan, *Seismic metamaterials for low-frequency mechanical wave attenuation*, Natural Hazards, 1-17 (2021)
7. A. A. Zadpoor, *Mechanical meta-materials*, Materials Horizons, 3(5), 371-381 (2016)
8. C. Z. Fan, Y. Gao, & J. P. Huang, *Shaped graded materials with an apparent negative thermal conductivity*, Applied Physics Letters, 92(25), 251907 (2008)
9. B. F. Spencer Jr, & S. Nagarajaiah, *State of the art of structural control*, Journal of structural engineering, 129(7), 845-856 (2003)
10. R. Xu, M. Muzamil, L. Fan, K. Yuan, H. Yang, & Q. Du, *Broadband seismic metamaterial with an improved cylinder by introducing plus-shaped structure*, EPL (Europhysics Letters), 133(3), 37001 (2021)
11. F. Basone, O. S. Bursi, F. Aloschi, & G. Fischbach, *Vibration mitigation of an MDoF system subjected to stochastic loading by means of hysteretic nonlinear locally resonant metamaterials*, Scientific Reports, 11(1), 1-15 (2021)
12. H. J. Xiang, Z. F. Shi, S. J. Wang, & Y. L. Mo, *Periodic materials-based vibration attenuation in layered foundations: experimental validation*, Smart Materials and Structures, 21(11), 112003 (2012)
13. T. V. Varma, B. Ungureanu, S. Sarkar, R. Craster, S. Guenneau, & S. Brûlé, *The influence of structure geometry and material on seismic metamaterial performance*, arXiv preprint arXiv:2007.14424 (2020)
14. B. Ungureanu, Y. Achaoui, S. Brule, S. Enoch, R. Craster, & S. Guenneau, *Seismic wave shield using cubic arrays of split-ball resonators*, arXiv preprint arXiv:1904.10767 (2019)
15. M. M. Hajjaj, & J. Tu, *A seismic metamaterial concept with very short resonators using depleted uranium*, Archive of Applied Mechanics, 1-22 (2021)
16. S. Dharmaraj, V. Ashokkumar, R. Pandiyan, H. S. H. Munawaroh, K. W. Chew, W. H. Chen, & C. Ngamcharussrivichai, *Pyrolysis: An effective technique for degradation of COVID-19 medical wastes*, Chemosphere, 275, 130092 (2021)
17. M. Aggarwal, *Pollution watchdog releases guidelines to handle COVID-19 biomedical waste*, Retrieved June, 30, 2020 (2020)
18. H. B. Sharma, K. R. Vanapalli, V. S. Cheela, V. P. Ranjan, A. K. Jaglan, B. Dubey, & J. Bhattacharya, *Challenges, opportunities, and innovations for effective solid waste management during and post COVID-19 pandemic*, Resources, conservation and recycling, 162, 105052 (2020)
19. S. Jung, S. Lee, X. Dou, & E. E. Kwon, *Valorization of disposable COVID-19 mask through the thermo-chemical process*, Chemical Engineering Journal, 405, 126658 (2021)
20. M. Miniaci, A. Krushynska, F. Bosia, & N. M. Pugno, *Large scale mechanical metamaterials as seismic shields*, New Journal of Physics, 18(8), 083041 (2016)
21. H. T. Özkahraman, & M. Şirin, *Göltaş Kireçtaşının Kayaç Kalitesi ve Öğürülebilirliğinin Tespiti*, Geosound, Number: 33, pp. 155-162 (1998)
22. M. Abu-Saleem, Y. Zhuge, R. Hassanli, M. Ellis, M. Rahman, & P. Levett, *Evaluation of concrete performance with different types of recycled plastic waste for kerb application*, Construction and Building Materials, 293, 123477 (2021)
23. H. Binici, M. H. Alma, R. Gemci, & M. Y. Durgun, *Atık Polietilen (PE) bardaklardan üretilen çimentosuz harçların fiziksel ve mekanik özellikleri*, Yapı Teknolojileri Elektronik Dergisi, 7(1), 71-79 (2011)

https://doi.org/10.1051/e3sconf/202340403003

24. A. V. C. I. Ahmet, & Ö. F. Erkendirci, *Farkli Çatlak Açilarina Sahip Çelik Tel Takviyeli Polietilen Kompozitlerin Yorulma Yükü Altinda Kirilma Davranışlarının İncelenmesi*, Prooceedings of 8th International Fracture Conference (2007)

Introduction of optimal photovoltaic and wind power balance algorithm in power systems

Emmanuel Karapidakis[1], *Marios* Nikologiannis[2], *George* Konstantinidis[1], and *Yiannis* Katsigiannis[1*]

[1] Hellenic Mediterranean University, Electrical and Computer Engineering Dept, 71410, Greece.
[2] University of Crete, Physics Department, 70013, Greece.

Abstract. Since 2021, Crete has been connected to the mainland's power system grid providing security regarding load peak demand and intra-day fluctuations. The grid provides, also, the opportunity to take full advantage of it in order to alter the island's energy mix, making the island's power system more climate neutral and reduce the emissions produced by conventional power production means. In the current paper, the utilization of mathematical models and a few numerical solution algorithms lead to the optimization of the power production of RES, under certain constraints regarding the system's rejections. The approximation of each optimal solution for all three cases is satisfactory, which is, for the first two cases, an immediate result of the linear properties of the objective function that expresses the annual production of RES in each case. Those linear properties prevent each approximated solution from deviating from the problem's feasible region, while, also, permitting the use of optimization techniques which can be easily implemented and require very little computational time. For the third case, a search algorithm has been designed and implemented, yielding accurate, yet low-precision solutions, which approximate sufficiently the optimal ones.

1 Introduction

For the past years, Crete's energy demands have been satisfied mostly with conventional means of production, as well as renewable energy sources (RES) [1]. However, the RES share was relatively low in the island's energy mix primarily due to their unpredictable output that is dependent on local weather conditions [2]. This causes significant reliability to conventional units that have a highly controllable output yet produce power by consuming fuels (such as mazut, natural gas, etc.) whose prices and means of supply are sensitive to geopolitics and macroeconomic events while contributing significantly to the local area's pollution levels.

The investigation of increasing the RES penetration in the power systems has been examined in the literature taking into consideration several points of view. Luz et al. [3] used the Interactive MOLP Explorer to solve the multi-objective linear problem of minimize the levelized cost of electricity and maximize the generation of non-conventional

* Corresponding author: katsigiannis@hmu.gr

RES. The results show that 90% of the annual load can be covered by renewable sources and that solar power can play a significant part in the power share. In the same topic from a different perspective, the authors in [4] estimated an increase of RES penetration with energy storage system at the point of 60-90% of the total energy of islands can be feasible a project. The authors used 'Response Optimization' tool within the Simulink model to minimize the levelized cost of the system. In [5] the authors estimated that up to 90% of the electricity demand of Dia island, which is located in Greece, can be supplied by RES. In [6] an increase of 65.5% of RES was estimated in a protentional interconnection between Pico and Faial islands. It is used the EnergyPLAN tool to evaluate the increase of RES and the installation of energy storage system to the power system of the two islands. A similar work is performed in [7], where the authors concluded that the interconnection of the Lesbos island with the mainland can increase the share of wind turbines and make the interconnection a more feasible scenario.

The interconnection of Crete's power grid connection with the mainland has provided an opportunity to redesign the current power system allowing higher RES penetration in the energy mix providing economic and environmental benefits [8]. This can be achieved by optimizing the penetration of photovoltaics and wind turbines under certain constraints regarding the rejection of energy by the system. The goal is to reduce the dependence on conventional means of power production, reduce the import of energy through the interconnection with the mainland, and turn the island into an energy exporter for certain hours of the year.

In this paper, mathematical models are designed in order to determine the functions that relate the nominal powers of the photovoltaics and wind turbines to be installed with the annual energy produced by RES, the annual amount of energy imported, and the annual amount of energy rejected by the system. Numerical solution algorithms are also used to determine the optimal nominal powers for each case.

2 Power system study

For the purpose of this study, data from the beginning of 2022 until the end of the same year regarding hourly demand (load), RES production, power import/export from the interconnection, production from the generators, as well as normalized power output of solar panel and wind turbine parks has been provided. For the year 2022, nearly 2/3 of the energy mix of the island was consisted of energy originating from conventional means of production, while production of RES was 23.16% and the contribution of the interconnection 12.80%. There are three distinct cases under study in this paper:

Case 1: Increasing RES by installing additional solar panels.

Case 2: Increasing RES by installing additional wind turbines.

Case 3: Increasing RES by installing additional solar panels and wind turbines.

In all cases the same constraint is enforced, which dictates that the percentage of rejections of the system does not exceed 10%.

3 Methodology

3.1 Case 1: Installation of additional solar panel nominal power

In this case, for each hour i the island's load (Load[i]) is satisfied with power production from RES (RES[i]), conventional generators (Conv[i]) and, if needed, from imports coming from the mainland's grid (Link[i]). If the production from RES and conventional generators suffices, then the surplus is exported through the interconnection to the mainland. Using mathematical notation, the above can be symbolized as:

$$Load[i] = Conv[i] + RES[i] \pm Link[i]. \quad (1)$$

With the increase in RES production (NewRES[i]), the new contribution of RES each hour i is:

$$NewRES[i] = RES[i] + x*PV_{norm}[i], \quad (2)$$

where x is the nominal power of the additional solar panels to be installed, and PVnorm[i] the normalized energy produced by the solar panels.

In an effort to reduce the system's dependency on conventional generators and locally-produced emissions, the RES output has been prioritized overall and the imports have been prioritized over fuel-consuming production. Which means that:

$$\pm NewLink[i] = Load[i] - NewRES[i] = Load[i] - RES[i] - x*PV_{norm}[i]. \quad (3)$$

By adding each term i for all 8760 hours of the year, the above equation becomes:

$$\sum_{i=1}^{8760} \pm NewLink[i] = \sum_{i=1}^{8760} Load[i] - \sum_{i=1}^{8760} RES[i] - x*\sum_{i=1}^{8760} PV_{norm}[i]. \quad (4)$$

The summed terms in equation (4) are the annual energy imported/exported to/from the system, the annual energy consumed by the power system's users and the annual energy produced by RES and conventional production systems.

In an ideal scenario the generators wouldn't be needed at all. However, there is a limit to the amount of power that can be imported/exported each hour, meaning that if the RES production and the maximum imported energy do not suffice, then the use of the generators is obligatory. With that in mind, each term NewLink[i] is calculated using the flowchart depicted in Figure 1:

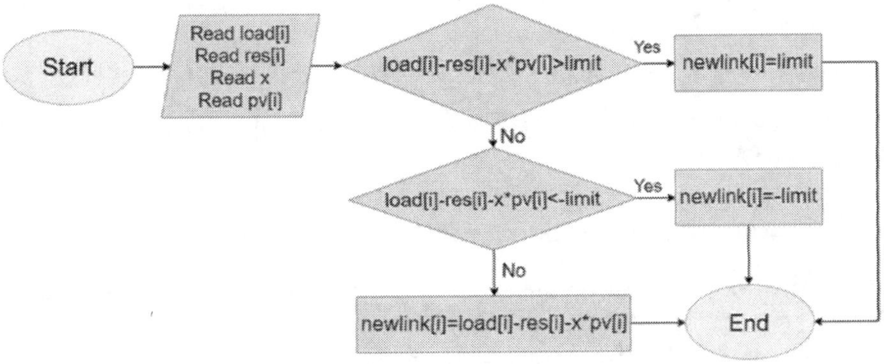

Fig. 1. Calculation of NewLink[i] flow chart.

The export limit presents a different problem. If the RES production is more than a specific hour's need (in this case during sunny hours), then the remaining is exported through the interconnection. If the amount of power to be exported exceeds the interconnection's capacity, then the rest is discarded completely. The discarded energy is called the rejection of the system and plays an important part in the optimization problem. The algorithm for the calculation of rejection Rej[i] is described in the following flow chart in Figure 2:

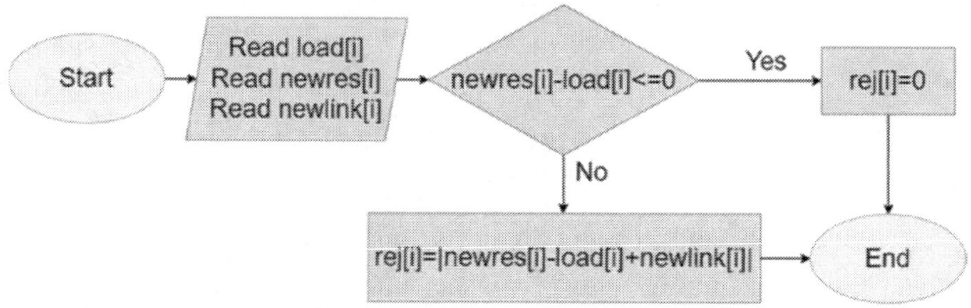

Fig. 2. Calculation of Rej[i] flow chart.

Using the flow charts in Figure 1 and Figure 2, the annual Rejections sumRej(x) is plotted and a polynomial function f(x) is fitted to them. Applying the constraint that annual rejections should not exceed 10% of energy produced by RES, which is described by the inequality:

$$\frac{sumRej(x)}{sumNewRES(x)} \leq 0.1, \tag{5}$$

where:

$$sumNewRES(x) = sumRES + x*sumPV_{norm}. \tag{6}$$

The optimal x value is being searched that maximizes sumNewRES(x). SumNewRES(x) increases linearly with x, so the optimal solution is expected to be found where the inequality becomes an equation, meaning g(x)=0.

Once the equation g(x) is determined, the root is calculated using the Newton-Raphson method. This method is best suited for finding the roots of differentiable, continuous functions. The method is applied in a range of the variable [a,b], usually with f(a)*f(b)<0. Starting from an initial point x0 ∈ [a, b], the value f (x0) is calculated and its first derivative $f'(x_0)$. The new approximation, x1, is the intersection point between y = 0 and the straight line determined by f'(x0), so $y=f'(x_0)*(x- x_1)$. For $x=x_0$, $f(x_0)=f'(x_0)*(x- x_0) \Rightarrow x_1=x_0-\frac{f(x_0)}{f'(x_0)}$. The method iterates until a value $x_n=x_{n-1}-\frac{f(x_{n-1})}{f'(x_{n-1})}$ is found, ideally with $f'(x_{n-1}) \neq 0 \; \forall \; x_{n-1}$ with $f(x_n) \leq \epsilon$, or until two consecutive x values, x_n, x_{n-1}, satisfy $|x_n-x_{n-1}| \leq \epsilon$, where ϵ the specified by the user error.

3.2 Case 2: Installation of additional wind turbine nominal power

Similar to the previous case, the optimal value y is searched, where y the nominal power of the wind turbines to be installed. The current model, in comparison to the previous one, has slight changes to the NewRES[i] formula:

$$NewRES[i] = RES[i] + y*WT_{norm}[i], \tag{7}$$

where WTnorm the normalized output of the wind turbines. The imports/exports from the interconnection are equal to:

$$\pm NewLink[i] = Load[i] - NewRES[i] = Load[i] - RES[i] - y*WT_{norm}[i]. \tag{8}$$

The algorithm described in Figure 1 changes slightly to the one in Figure 3:

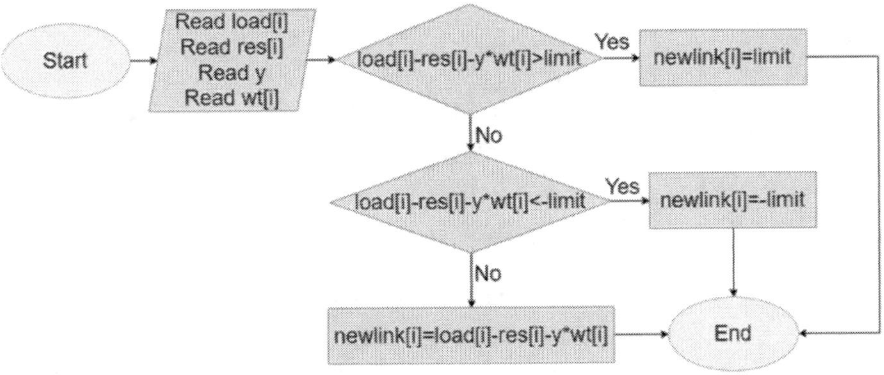

Fig. 3. Calculation of NewLink[i] flow chart.

The algorithm described in Figure 2 remains the same, the Newton Raphson method that maximizes sumNewRES under the constraint described in Case 1 is, once again, used to determine the optimal value of y that satisfies g(y) = 0.

3.3 Case 3: Installation of additional solar panel and wind turbine nominal powers

In this case, both the solar panel and wind turbine parks are reinforced, resulting in the NewRES[i] formula:

$$NewRES[i]=RES[i]+x*PV_{norm}[i]+y*WT_{norm}[i]. \quad (9)$$

The algorithm for the calculation of NewLink[i] changes once more to Figure 4:

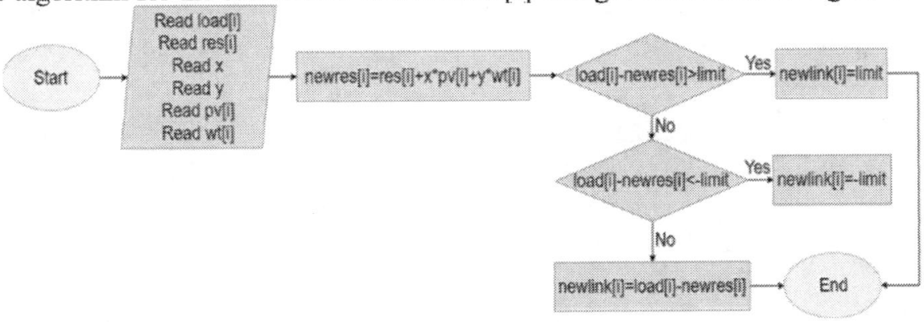

Fig. 4. Calculation of NewLink[i] flow chart.

While the algorithm for the rejections remains the same. Using those algorithms, the percentage of the amount of rejections from the amount of energy produced by new RES is calculated for x in range [0,1650] and y in [0,1800] and a contour map of those values is created and depicted in Figure 5:

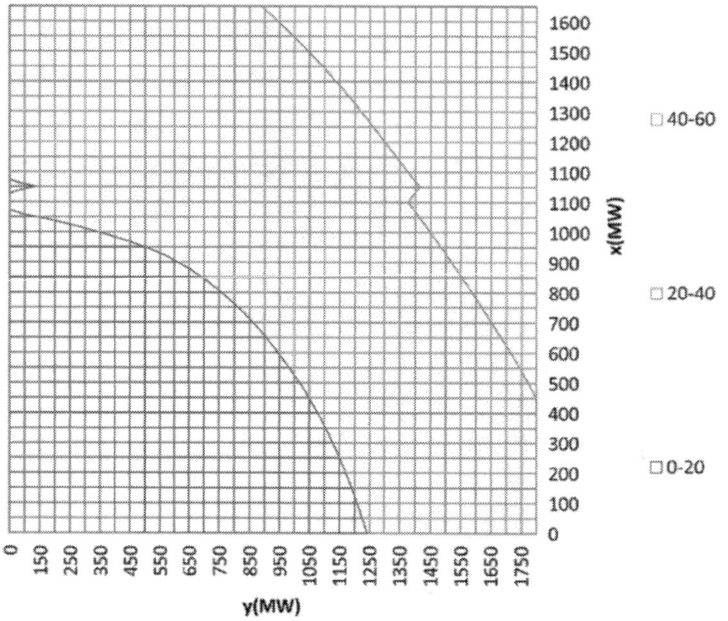

Fig. 5. Contour map of sumRej(%) for link limit=150MW per hour.

In Figure 5, it is apparent that the (x,y) pair that assures maximum RES energy production with rejections less than 10% of the produced energy is located in [0,1250] for both x and y.

For the location of the wanted x and y, a searching algorithm is implemented which reads all the hourly values of load, res, pv and wt from a .txt file, all the (x,y) possible combinations are tested in order to locate the pair that insures the maximum RES production under the forementioned constraint. The first three subroutines are presented in Figure 6.a, and the rest in Figure 6.b:

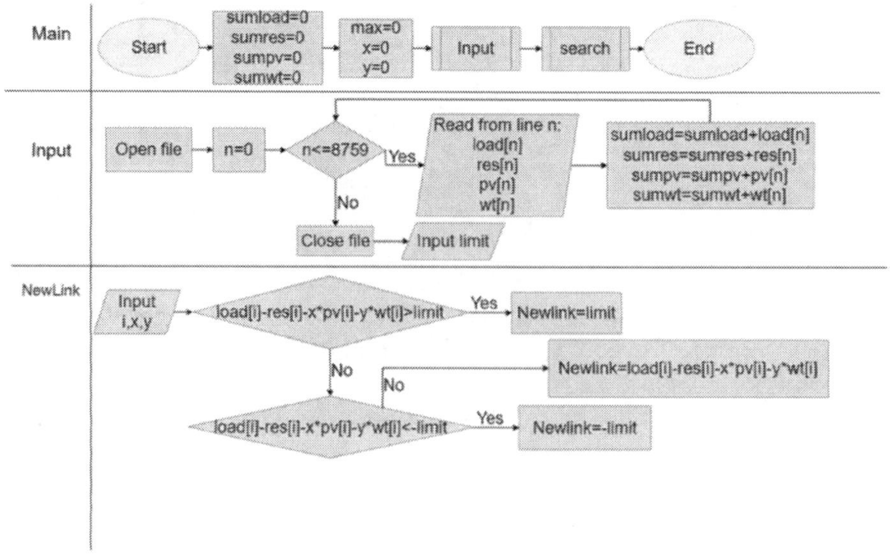

a)

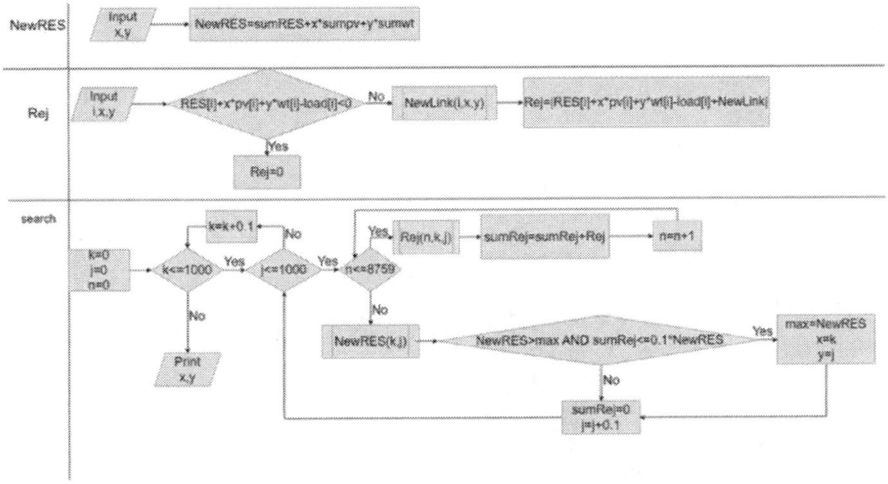

b)

Fig. 6. Optimal values search algorithm for x and y of scale 1 flow chart.

4 Results

4.1 Case 1

In the first case, the relation between the solar panel nominal power and the annual amount of energy that is rejected is presented in Figure 7:

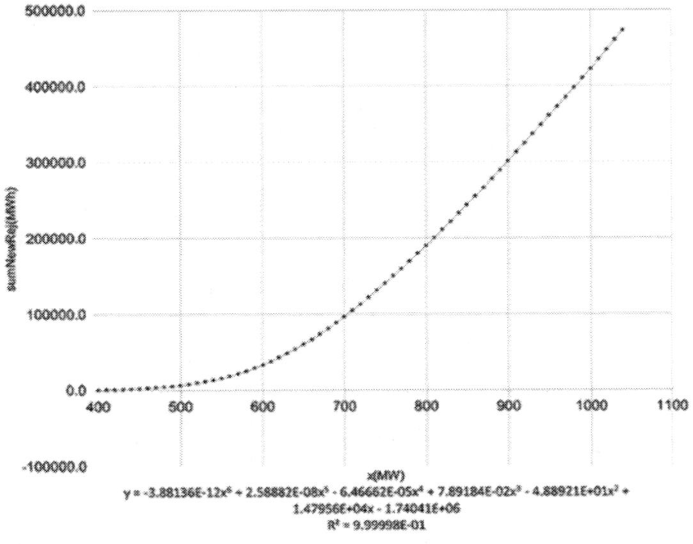

Fig. 7. The sumRej(x) plot for link limit=150MW per hour.

The constraint from equation (5), therefore, turns into:

$$-3.88136*10^{-12}*x^6+2.58882*10^{-8}*x^5-0.0000646662*x^4+0.0789184*x^3-48.8921*x^2 +14625.2*x-1.81528*10^6 \leq 0. \quad (10)$$

Applying the Newton-Raphson method in the range [0,850] with $\epsilon = 10{-}10$, the optimal solution found is x = 824.126MW with $g(x) = -1.6415 * 10^{-9}$MWh. Applying this solution x for link limit=150MW per hour and the same load, the energy mix would be as presented in Figure 8:

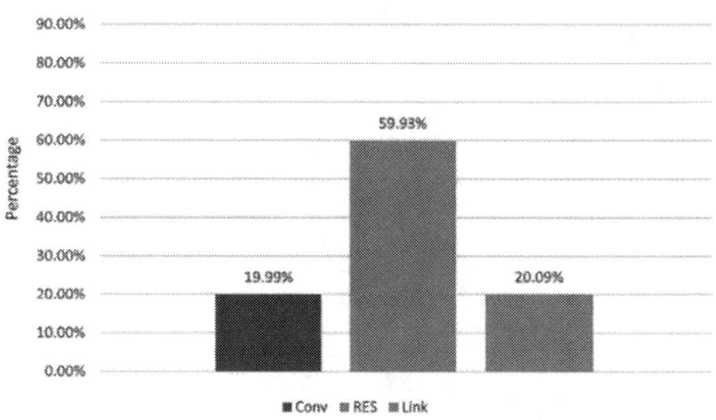

Fig. 8. Energy mix for optimal x and annual Rejection percentage equal to 10.0022%.

It is apparent from Fig. 8 that installing extra nominal power of photovoltaic parks alone increases RES penetration to a little less than 60%, with a significant reliance on both conventional means of power production and imports through the interconnection, with a total annual amount equal to 40.08% of annual load demand.

4.2 Case 2

Similarly, for the newly installed wind turbines, the plot that presents the annual rejections of the system related to y is shown in Figure 9:

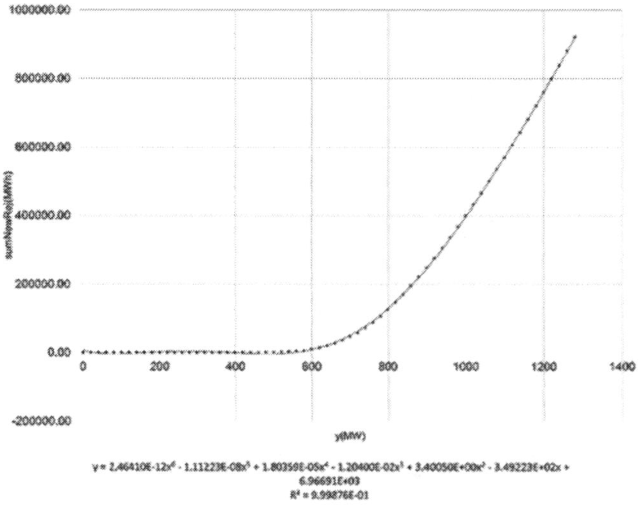

Fig. 9. The sumRej(y) plot for link limit=150MW per hour.

Now, the constraint from equation (5) becomes:

$$2.4641*10^{-12}*y^6-1.11223*10^{-8}*y^5+0.0000180359*y^4-0.01204*y^3 +3.40050*y^2-628.14*y-67899.7\leq 0. \quad (11)$$

Applying the Newton-Raphson method once more in the range [500,1000] with $\in=10^{-10}$, the optimal nominal power for wind turbines is y = 968.266MW with $g(y) = -1.44064 * 10^{-9}$MWh. Applying this solution for link limit=150MW per hour, the energy mix would be as depicted in Figure 10:

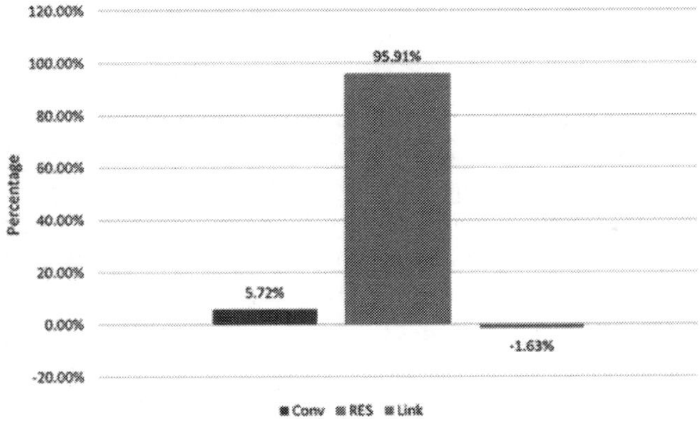

Fig. 10. Energy mix for optimal y and annual Rejection percentage equal to 10.1010%.

In contrast to the previous case, the installation of added wind turbine park's nominal power, as can be seen in Figure 10, increases greatly the annual RES production, while the reliance to greenhouse gas emissions-producing means decreases to considerably low levels of 5.72% of annual demand. It should, also, be noted that the annual exports through the interconnection exceed the annual imports, contributing to the system's self-reliance.

4.3 Case 3

Implementing the algorithm from Figure 6, the optimal solution is x = 345.6MW for the solar panels and y = 808.1MW for the wind turbines. Applying these findings, for link limit=150MW per hour, the energy mix would be as shown in Figure 11:

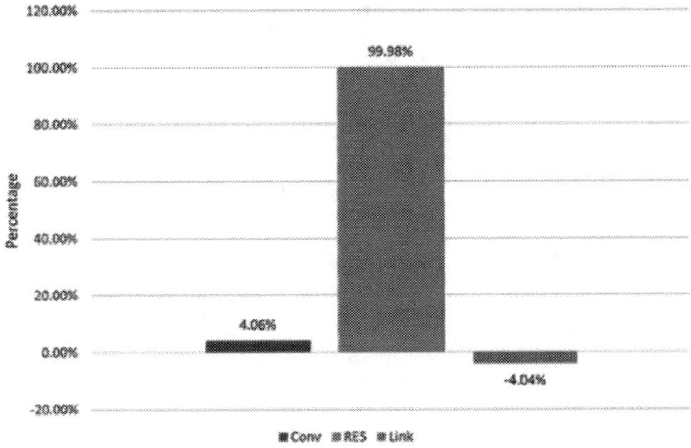

Fig. 11. Energy mix for optimal x and annual Rejection percentage equal to 9.99885%.

From a technical perspective, the search algorithm lacks precision of the approximated solutions, yet it is quite accurate, yielding annual rejection percentage of the system close to the 10% which the optimal solution provides. For this output of the algorithm, Figure 11 clearly dictates that the production of RES reaches nearly 100% of the annual system's demand, with a direct result being the low levels of dependency on fuel-consuming production methods, and the increase of the annual exports through the interconnection to 4.04% of the annual island's demand.

5 Conclusions

The goal of this paper was to study the energy system of Crete as an example of an interconnected island system in order to optimally maximize the penetration of RES energy production in the system's energy mix. This was pursued by designing a mathematical model for three distinct cases and use numerical analysis algorithms to determine the optimal solution for each case under a specific constraint regarding the energy rejections of the system (≤10%). The first case was the installation of extra solar panel nominal power which achieved additional 36.77% penetration of RES to the mix, the second case was the installation of extra wind turbine nominal power to the system, resulting in 72.75% increase in RES penetration to the energy mix, and the final case of increasing both solar panel and wind turbine nominal power that increases RES penetration by 76.82%. It is noteworthy that in the last case the contribution of RES surpasses 99%, since both power-production technologies are taken advantage of, enabling the system to produce power originating from renewable sources for more hours of each day.

At this point, we would like to acknowledge that the research project was supported by the Hellenic Foundation for Research and Innovation (H.F.R.I.) under the "2nd Call for H.F.R.I. Research Projects to support Faculty Members & Researchers" (Project Number: 03698).

References

1. A. G. Tsikalakis, Y. A. Katsigiannis, E. S. Karapidakis & K. E. Fiorentzis, *Evaluating the effect of wind-hydro hybrid power stations on the operation of Cretan power system*, in Proceedings of 52nd International Universities Power Engineering Conference (UPEC), 28-31 August 2017, Heraklion, Greece (2017)
2. K. Fiorentzis, A. Tsikalakis, E. Karapidakis, Y. Katsigiannis & G. Stavrakakis, En. **13**(1), 64 (2019)
3. T. Luz, P. Moura, A. de Almeida, Ren. and Sust. En. Rev. **81**(2), 2637-2643 (2018)
4. D. M. Gioutsos, K. Blok, L. van Velzen, S. Moorman, Appl. En. **226**, 437-449 (2018)
5. D. Al. Katsaprakakis, N. Papadakis, G. Kozirakis, Y. Minadakis, D. Christakis, K. Kondaxakis, Appl. En. **86**(4), 516-527 (2009)
6. M. Alves, R. Segurado, M. Costa, En. **182**, 502-510 (2019)
7. M. Kapsali, J.K. Kaldellis, J.S. Anagnostopoulos, Appl. En. **173**, 238-254 (2016)
8. P. N. Geogiou, G. Mavrotas, D. Diakoulaki, Ren. and Sust. En. Rev. **5**(6), 2607-2620 (2011)

Study on behaviour of centrifugal pump driven by medium-voltage induction motor during operation control

Svilen Rachev[1*], *Kremena* Dimitrova[2], and *Lyubomir* Dimitrov[1]

[1]Technical University - Gabrovo, Faculty of Electrical Engineering and Electronics, 5300 Gabrovo, Bulgaria
[2]Technical University - Sofia, Faculty of Engineering and Pedagogy, 8800 Sliven, Bulgaria

Abstract. When studying electric drives, it is necessary to jointly analyze the processes in the electric power and control circuits of the electric motor as the main component determining the overall performance and in the mechanical executive part. The paper deals with the variation of the occurring torques, the currents flowing in the individual phases and the electrical power losses in the operation of a medium voltage electric drive of a pump unit from the point of view of energy efficiency. Emphasis is placed on the joint operation of the pump as a mechanism and the driving electric motor, studying the complete electromechanical system. During the studies, significant attention has been paid to the mathematical description and modeling with subsequent computer simulation of the ongoing electromagnetic and electromechanical processes. Different ways of controlling the motor have been applied - by softstarter and by frequency control. Also values for the duration of the start-up process have been obtained. Relevant conclusions have been drawn.

1 Introduction

Electric motors (EMs) are a dominant class of electrical energy users in the world. Therefore, research in the field of minimization of the energy consumption of electric drives is always relevant.

Considering the position of EMs as one of the largest consumers of electricity, even small improvements in their efficiency can lead to significant energy savings, which will also contribute to increasing sustainability worldwide.

When designing EMs, the goal is to achieve maximum efficiency at 60÷80% of the rated load, since they usually operate with an under-load of 15÷25%. To move the maximum efficiency in the rated load area or in the overload area, the cross-section of the windings should be increased and the electrical losses in the EM should be reduced.

The induction motor (IM) with a squirrel-cage rotor is the most commonly used rotating electric machine. The reason for this is the simple design of this type of EM, which leads to:

* Corresponding author: sratchev@mail.com

- low price of the rotating machine.
- cheap service (no need to check and replace brushes, nor clean the dust compartments of the brushes.
- high reliability.

The large starting current is a major disadvantage of the IM in direct on line starting (DOL starting). The quadratic dependence of the IM torque on the supply voltage is its other main disadvantage. Short-term dips in supply voltage can reduce the motor's torque to such an extent that it stops rotating altogether. Voltage dips represent a short-term but significant reduction in the effective value of the voltage in one or more phases. There are two main causes of voltage dips in the power system – commutations in the electrical energy system and poor connections, mostly between aluminium cables and copper cable lugs, leading to a potential contact difference.

It is known that pumps are working machines that serve to suck a liquid standing at a lower level and push it to the desired height to a higher level. To carry out this process, work is expended, produced in other machines, called *power machines*, such as water and steam turbines, EMs, and the like. Pumps are responsible for about 8-9% of the global consumption of electric energy. Large electric motors with more than 375 kW nominal output power are usually high-voltage AC motors and they account for about 23% of all electric energy consumption by motors. [1].

A certain number of pumping stations with pumping units operating in them are always provided for each populated place with a view to drinking water supply. Very often 1, 2 or more pump units are in operation, and even more are available as spares, and these are most often repaired with rewound coils. In most cases, the drive IMs are powered at medium voltage (MV), most often 6 kV and not at low voltage (LV) 400 V.

The higher the voltage level, the higher the installation costs; although the current as the voltage increases becomes smaller, the wire and cable can be selected with smaller sections, but the cost of other medium-voltage circuit breakers, transformers, switchgear and other equipment still increases. The initial investment is large, so small businesses are ready to use LV equipment when they are newly registered, i.e. startups.

The higher the voltage level, the lower the total operating costs; the small current leads to a reduction in power losses, which is suitable in the long run. The cumulative effect is incredible, as many large enterprises replace LV motors with MV motors during technical transformations.

Based on economic calculations, the main costs at pumping stations are for electricity. With the continuous increase in the price, it turns out that about 40% of the energy consumed is practically wasted. Therefore, as an energy-efficient solution with a great economic effect, control of EMs for pump units can be used.

The traditional water supply system, without regulation of the IM drive, can only provide a constant pressure regardless of consumption or momentary loads. The addition of a frequency regulator enables the water supply system to adjust the water pressure as required. When there is no demand for water, the frequency control goes into "sleep" mode until water is needed again and then immediately provides constant water flow. The effect of this control method is to reduce the electrical energy costs of pumping units by 10 to 30%. This results in significant water and electricity savings, which is particularly noticeable in environments with inconsistent water flow. The frequency electric drive reduces the average operating speed of the pumping unit to extend the life cycle and adjusts the output speed according to the needs.

Very often, large currents are obtained during transient processes. They usually decay very quickly, but their maximum values are very large. The amplitude of the current through the transient process is called the *impact current*. It can exceed as a value the starting current declared by the manufacturer as a multiplicity of the rated one. It is necessary to mention that modern electronic static electricity meters (without rotating parts) are very precise and take into account the large consumption of electrical energy when starting powerful electric drives.

Another consideration must be kept in mind. When large MV motors or combinations of MV motors are started, large instantaneous amounts of reactive power are required. This depletion of reactive power from the distribution system can cause objectionable disturbances on the system. Starting a motor with full voltage imposes a high reactive power requirement, with a low, lagging power factor, for a short period of time, during its acceleration period. [2]

The mechanical effort in the windings is proportional to the current squared, and therefore the need to determine the currents during the transient processes, and hence the knowledge of the process itself, is clear. Since the impact current is many times greater than the rated one, it is clear that there will be forces acting mostly on the butt joints of the windings, which have large values.

Large values of torques can also be obtained during transient processes. In this case, the maximum value of the torque can exceed the rated one several times. This maximum torque acts as a shock on the EM and on its shaft, i.e. on the entire kinematic scheme of the electric drive, questioning its reliable operation – hence the name *impact torque*.

Transient processes in many cases are the cause of deterioration of the quality of electrical energy. The main voltage parameters and their permissible deviation ranges are regulated in [3]. In general, voltage dips are a problem affecting the electrical qualities [4].

All of the above substantiates the need to search for technical solutions ensuring the normal operation of pumping units, while at the same time the focus is on the universal orientation towards energy efficiency. Only the technical side of the issue is considered without analyzing the economic benefits of the implementation of energy saving methods.

In brief, the paper deals with the determination of the torques occurring, the stator currents flowing and the electrical power losses in the operation of a medium voltage electric drive of a pump unit from the point of view of energy efficiency, applying two different control methods - a soft starter and a frequency inverter.

2 Methodology

Starting, stopping, maintaining and changing the speed of EMs are basic tasks of industrial automation. All starting and control methods available for MV induction motors are considered successively in [2]. In terms of means of control, EMs can be classified into several categories – schemes without control and those with control.

Schemes without control. Typical connection schemes provide start, stop, short circuit protection, thermal protection, zero protection. Thermal protection also indirectly protects the EM from two-phase operation and under-voltage operation, but in some cases special protections can be added to prevent these modes. Zero protection is self-start protection – if the supply voltage fails during operation, the EM will stop, and when the power is restored, the circuit will not allow the EM to self-start and will wait for an operator to give a start signal.

The self-starting of the EMs is dangerous both from a mechanical (for the safety of the people working with the mechanism) and from an electrical point of view – if a large number of EMs are connected to the same busbars, in the event of a self-starting, they will start spinning at the same time, their large starting currents flow, which can cause a serious voltage dip in the direction of falling well below the rated level. Low voltage will aggravate the already difficult starting process and may lead to starting problems or unacceptably long running of high starting currents, and therefore overheating of the EMs.

In DOL starting, the EM is connected to rated voltage at rated frequency. At the first moment of time after launch, the rotor will still be stationary due to its inertia. By analogy with a transformer, if we are considering a squirrel-cage rotor IM, the secondary winding will be short-circuited. This explains the large starting current that flows then – up to 7 times the rated current of the EM. DOL starting is simple and inexpensive to implement and is used whenever the drive mechanism, thermal considerations and the influence of the inrush current on adjacent loads permit. DOL starting is used when providing stable power to a EM that is hardwired to the drive, such as a pump. The main considerations that limit the application of DOL starting are:

- The large starting current causes intense heating of the EM. The large starting current causes significant thermal overloads on the winding, which can lead to accelerated aging of the insulation, damage to it and, as a result, a short circuit between windings. In addition, it is accompanied by large electro-dynamic forces acting on the EM windings. Due to the uneven distribution of resistances, the large starting current can also cause over-voltages between the windings of the stator coil.
- The starting current of powerful EMs or when starting many EMs at the same time causes a strong voltage dip on the supply busbars, which affects other consumers and the EMs themselves, further complicating the starting process [1]. The power supply transformers are also overloaded.
- Another disadvantage is the fluctuating damping nature of the motor's starting torque. Significant fluctuations in the EM torque in the initial stage of start-up, which can exceed the rated torque many times over, create unfavourable conditions for the operation of the mechanics.

It is possible in rare but not unusual modes to experience a torque deficiency in some mechanisms, i.e. the starting torque is less than the resisting moment of the mechanism and the EM does not spin at all. This is effectively short-circuiting the EM. Because of the listed factors, in more specific applications it is necessary to control the starting process and avoid DOL starting.

Schemes with process control. Various schemes are applied to reduce the large impact current and limit the impact torque, for example, in mechanisms with a fan static resistance characteristic, in which during rotation, i.e. at low rotational frequencies, the resisting moment is small and increases with the second (or greater) power of the rotational frequencies. A possible technical solution is the use of softstarters, providing smooth starting of the EMs, which can also provide some protections.

In softstarters, digital control is integrated with electromechanical starters, which enables smooth acceleration of the EM both to prevent damage to transmissions and to avoid pressure on the power distribution company when starting medium and large EMs at full voltage. Under the general name softstarters, there are many technical developments that differ from each other in terms of their capabilities, connection schemes, need for additional devices.

The power circuit of commercially available softstarters is built as a thyristor (some sources use SCR as a synonym – silicon controlled rectifier) regulator with phase control. By adjusting the opening angle of the thyristors, the effective value of the voltage applied to the EM is controlled. Since they are built with thyristors, which are among the most reliable semiconductor switches, softstarters are characterized by simple construction and operational reliability.

A significant proportion of softstarters could be divided into the following four categories:

- torque limiters.
- regulators without feedback.
- regulators with feedback and current limiting.

- regulators with feedback and current control.

Softstarters provide smooth cranking up to full speed and are used when starting (and stopping if equipped). Soft start is achieved by gradually applying the voltage to the EM using thyristor switches connected in series in the supply line of the EM or in series on each of the windings connected in a delta.

They are often used in pumps to eliminate sudden pressure rises in pipes when the fluid flow rapidly changes direction. In some hydraulic systems with centrifugal pumps, a sudden stop of the pump can cause hydraulic shock. In these cases, softstarters can be used to smoothly reduce the supply voltage and hence the EM torque.

The voltage supplied to the EM is reduced by changing the opening angle of the thyristors – symmetrically for the positive and negative half-waves of the voltage. There are simplified circuits with a single thyristor and reverse diode, but these are not recommended because they are unbalanced and generate even harmonics. The thyristors operate during ramp-up, and after reaching rated speed they are shunted by bypass switches. This significantly reduces EM heating.

Softstarters are usually the cheaper choice in cases where speed and torque are required only when EM starts. In addition, they are very suitable for limited spaces, as they usually take up less space than frequency regulators. Softstarters can reach efficiency up to 99.5÷99.9%. Typically, the voltage drop across the thyristor is less than 1 V. The efficiency depends on the size of the softstarter and the supplied three-phase voltage. After the start-up process is complete, the softstarter turns on the built-in bypass contactors. Thus, the thyristors are shunted and all the current flows through the contactors.

The following electronic element base is used: thyristors – the widely used ones can switch up to 1200 V (leading developments up to 8000 V) and 4000 A. Modern modified variants of thyristors are those with disconnection from the control electrode GTO (gate turn off), GCT (gate commutated thyristor) and SGCT (symmetrical gate commutated thyristor), IGCT (integrated gate commutated thyristor) – GCT with a built-in control circuit. They can switch up to 4500 V (leading developments up to 6500 V).

Medium voltage softstarters usually use two or more thyristors connected in series in each arm, so that the mains voltage is distributed between them and cheaper thyristors with a lower voltage rating can be used.

Advantages of using softstarters for pump units:

- reduced hydraulic shocks in pipes during start and stop.
- the mechanical stress on the EM shaft is minimized.
- reduced starting current.
- low current protection prevents damage due to blocked pipe or low water level.
- automatic restart function ensures uninterrupted operation of a stand-alone pumping station.
- phase reversal protection prevents damage due to reverse pump rotation.
- instantaneous overload protection prevents damage due to garbage entering the pump.

Softstarters are load dependent and the start and stop settings are programmed. An algorithm adjusts the voltage so that the current and torque increase and the EM starts. Based on the reverse electromotive force of the EM, the softstarter determines whether it has reached the required rotation speed. If the softstarter detects that the EM has reached this speed before the scheduled start time, it supplies full voltage and turns on a normal operating status indicator. If the EM does not spin up to the required speed within the set time, the softstarter supplies the full voltage or part of it depending on the load.

For the longer cables/wires used to connect the softstarters, no special measures need to be taken, except that they should be sized correctly to compensate for the voltage drop. With softstarters, special measures and means of electromagnetic compatibility (suppression of the effects of electromagnetic influences) are usually not necessary, so as to satisfy the harmonics requirements according to International Electrotechnical Commission (IEC) standards.

As for any non-linear distortion, the softstarters generate a moderate amount of it – with the 5th harmonic being the biggest contributor. The percentage content of harmonics strongly depends on the degree of voltage reduction, i.e. is largest at the initial launch torque and decreases in the rollout process. The applicable standards tolerate short-term generation of harmonics, so no special measures are required in this regard. As for disturbances in the operation of surrounding facilities, in almost all cases the use of additional filtering devices is not required, but measures such as:

- moving the softstarter electrically further away from the power supply, respectively – closer to the EM.
- if possible – laying the power cables to and from the softstarter along separate routes from those of other power and control cables.
- use of shielded cables.

When choosing a softstarter, it should be taken into account that there are models with the additional possibility of directly connecting the EMs to the full mains voltage (DOL starting), which expands the applications.

The purpose of controlling EMs is to ensure the rotation of the rotor at a precisely defined speed, or to ensure its rotation at a precisely defined angle. The latter is called position control and is a more complex task. In speed control, the task is to apply such control effects to the EM that it maintains the set speed as accurately as possible with a resisting moment varying from zero to rated one. At the same time, it is known that if the powerful EMs work without load in an steady state mode, this means working with poor energy indicators – low efficiency and power factor.

In electrical engineering, the concept of frequency control is widespread. By itself, the term is correct, but not unambiguous. There are two major classes of approaches in the control of frequency inverters. The greatest opportunities are offered by the so-called *vector control*, in which very high accuracy is obtained. Vector control of IMs is truly an innovative method in energy saving. However, in order to implement it, a very good knowledge of the parameters of the specific EM is required. The second major class of approaches is called *scalar control*. With it, the inverter is controlled so that when the frequency changes, the voltage amplitude changes according to a predetermined law. The accuracy provided is significantly less, but the settings of the electronics-motor system are significantly easier. Furthermore, most industrial mechanisms do not require the vast capabilities that vector control offers.

In vector control, by means of mathematical transformations, a separation of the components of the supply three-phase quantities (voltage and currents) is achieved in such a way as to provide an analogy with DC motors – separate components are responsible for the magnetic flux and separate ones for the flowing current, and their product determines the shaft torque. There also appears the possibility of making current and flux perpendicular to each other. The torque is maximized when the flux linkage and current vectors are perpendicular [5]. The ideal scenario, in which field and current vectors are controlled to be orthogonal, is called *field orientation*. Field orientation, creates an ideal dynamic behavior just like the DC commutator machine [6].

The mathematical transformations allowing the separation of the components are based on the space vector theory, which by the way also gave the name of vector control. A necessary condition for achieving maximum efficiency. is the optimal choice of magnetizing current. Improper control of the magnetizing current can cause the consumption of a frequency-controlled EM to be more than that of a direct supply.

Since the basis of the drive's operation is to vary the frequency to the EM in order to vary the speed, the best-suited name for the system is the variable frequency drive (VFD).

However, other names used to reference this type of drive include adjustable speed drive (ASD), adjustable frequency drive (AFD), variable-speed drive (VSD), and frequency converter (FC) [7].

In speed control by changing the frequency of the power supply, the EM is connected to the mains via a frequency converter. Using this method, the important relationship of three parameters must be taken into account: phase voltage, mains frequency and basic magnetic flux. Determining here is the value of the magnetic flux, a significant change of which is undesirable: for example, with an increase in the flux, the saturation of the magnetic circuit increases and the magnetizing current increases sharply, and with a decrease in the flux, the rotor current can increase and it will overheat. Therefore, in most cases, magnetic flux $\Phi =$ const. should be maintained. It follows that simultaneously with the change in the frequency of the power supply, the voltage must also change proportionally. On the other hand, it should be noted that the law of voltage change depends on the characteristics of the working mechanism connected to the EM shaft.

When controlling an IM by varying the frequency and amplitude of the supply voltage, in order to keep the maximum torque constant, it is necessary to maintain U2/f = const. It can be considered that the slip remains unchanged if the resisting moment remains unchanged. Then the frequency of rotation of the rotor will be proportional to the frequency of the supply voltage. If the dependence of the inductive resistances of the windings on the frequency of the supply network is taken into account, in order to keep the maximum torque constant, it is necessary to maintain U/f = const.

If the active resistance of the stator winding is neglected, then the condition for a constant ratio of voltage and frequency would also mean an unchanged magnetic flux in the air gap, which is a condition for a constant torque. The latter law is the most applied governing law, because suggests a linear relationship between voltage and frequency.

It is important to mention that the bulk of IMs are produced with self-ventilation – fan blades are mounted on the end of the fan shaft opposite the end for connecting a mechanism or gear. Thus, when the EM rotates, it is blown, and therefore cooled, by its own fan. When the EM speed is reduced, the speed of the own fan also decreases, i.e. cooling worsens. For this reason, even with a current lower than the rated one, the EM may overheat. Therefore, where possible, the voltage is further reduced in order to reduce the current at low rotation frequencies, thereby reducing the heating of the EM. Therefore, when controlling centrifugal pumps and fans, the law U/f^2 = const. is sometimes used, in which the torque decreases strongly with a decrease in revolutions.

It is possible to increase the frequency of the supply voltage above the rated, but then the voltage cannot increase in proportion to the frequency, because the insulation of the EM windings will not withstand the increased voltage. So for over-synchronous speeds, only the frequency increases and the voltage remains equal to the rated one. Of course, this leads to a reduction in EM torque.

Operation at increased frequency is not so dangerous for the EM from an electrical point of view – there are increased losses in the steel and in the windings due to the skin effect. The high speed, however, places increased demands from a mechanical point of view. The higher the speed, the greater the centrifugal forces. Therefore, at over-synchronous speeds, special attention must be paid to the balancing of the rotor and the mechanisms connected to it. Eccentricities (in the sense of unbalanced masses) will result in strong knocking which can damage the bearings. Therefore, it must be certain that the selected EM will mechanically withstand the work at increased revolutions.

The main application area of scalar control of IMs is the drives of pumps, fans and work machines, where a variable speed is required to regulate the performance, but not very accurate job execution is required. Typical accuracy in open coupling systems is around 5% and improves when feedback is included, and can reach 0.5%.

Depending on the settings of the frequency converter, when low voltage is applied, a given pump may run at low speed. With low water consumption, running the pump at reduced power saves energy and increases EM life. But the most important thing is that at the moment of starting the pump, the EM starts to work at the smallest frequency, gradually accelerating to the set speed, which eliminates hydraulic shocks.

Thanks to the optimal control of the EM depending on the load, the consumption of electricity in pumps, fans, compressors and other units is reduced by 40÷50%, and the starting currents, which are 600÷700% of the rated ones and are a scourge for the equipment for starting and control, disappear completely. Thus, the use of adjustable electric drives based on frequency converters allows to create a new energy-saving technology that not only saves electrical energy, but also increases the service life of EMs and technological equipment in general.

Both a frequency regulator and a softstarter can be used to start and stop pumps. A softstarter reduces the effect of hydraulic shock when starting and stopping and is usually less expensive. A frequency regulator can do the same job, and in addition, it can also control the speed of the pump EM, respectively its flow rate, during operation.

Softstarters and electronic speed regulators have integrated protections. Modern speed regulators guarantee comprehensive EM overload protection as well as protection for themselves. Based on the current measurement and speed information, a microprocessor calculates the motor's temperature rise and issues an alarm or disconnect signal in the event of excessive overheating.

In addition, the information generated by the thermal protection integrated in the speed controller can be transmitted to a programmable logic controller or control system via the bus interface that newer speed controllers and starters have.

Some softstarters have limited ability to adjust the low speeds when starting and stopping. For example, for an EM with 1800 rpm possible low speeds are 18÷270 rpm. forward and backward. Maintaining low speeds is possible within a matter of minutes due to the temperature rise in the thyristors and in the EM.

Frequency converters provide continuous and complete speed control at all times, from start to stop, with the ability to maintain a given speed for hours. This is because the frequency is adjustable. Although both the frequency converter and the softstarter can operate at low speeds, the length of time they can be maintained is determined by the EM and the load. The degree of EM heating, if running at low speed, depends on the duration of operation. In order not to let it run too long at low speed, a temperature limit is provided in the softstarter – the aim is to protect the thyristors and the EM. Continuous operation of the frequency regulator at a frequency below 5 Hz imposes limitations on the permissible operating conditions.

Modern AC motor control systems work with electronic converters. To supply a EM with a voltage of variable frequency and variable amplitude, a rectifier-inverter system is used. By appropriate control of the inverter, a variation in both the frequency and the amplitude of the output voltage supplied to the EM can be provided.

Energy efficiency is a very important indicator that users of products and systems care about, and suppliers are working hard to improve it in their product range. In fact, according to the generally accepted opinion, the investment in the purchase of electrical equipment, as well as the costs due to the stoppage of the technological cycle in connection with its installation and commissioning, are compensated by the reduced energy consumption as a result of more energy efficient operation.

Electric motors often need large amounts of energy when accelerating rapidly to full speed. Both softstarters and frequency converters can be used to reduce impact currents and limit torque – thus protecting expensive equipment and extending EM life by reducing heating during frequent starts and stops. The choice between a softstarter and a frequency converter depends on the mode of application, the system requirements and the costs (both for purchase and implementation, and during the entire service life).

Mechanical shocks during start-up and load pulsations must also be considered. Any sudden change in torque from the EM side, or from the load side, leads to mechanical loads and fatigue in the shafts. In pumping units, sudden changes in torque can lead to hydraulic shocks in the system. This is one of the factors, along with thermal overload and increased energy losses during frequent starting (S4 mode), to favor soft-start methods, using a softstarter or frequency converter.

Directly related to the operation of rotating electric machines is the application of mathematical modelling, which plays a very important role in the design, exploitation and control of electric drives. Modelling and computer simulation, whether with regard to electric drive or in other branches of engineering, that is adequate and effective reduces the time needed and the cost of gaining an optimum design of a drive and its control system [8].

There are many definitions for optimization from the viewpoints of mathematics, science, and economics. A concise definition is given in [9]: "Optimization is the act to obtain the best results under given circumstance." For EM, the optimization could be applied in the machine (drive) design process in order to obtain the best machine (drive) performance according to a given criteria with minimum production cost, or it can be used in the drive control [10].

As pump requirements must match system characteristics, analysis of the overall system is necessary to establish pump conditions [11]. At best, the steady-state torque–speed characteristics could approximate the average of this dynamic response; it could not predict the complete dynamics during normal load torque changes for the larger motors [5].

When considering the dynamics of the electric drive, it is not possible to work with the effective values of the quantities, which is the approach when deriving the static characteristics. The processes in the electric drive in this case are described by differential equations. It is believed that electromagnetic processes lead to the appearance of an electromagnetic torque, which drives the shaft of the IM and, accordingly, initiates a mechanical transient process.

The startup properties of IMs are characterized by the initial startup and maximum torque, as well as the initial startup current. In IMs with squirrel-cage rotors, the torques and the startup currents depend on the ratio of the parameters. The parameters are the active and inductive resistances of the stator and rotor windings, the mutual inductance resistance and the moment of inertia of the rotor.

Very often, studies of electric drives with IMs are limited only to the determination of the initial startup and maximum torque and the initial startup current. This approach gives only approximate information about the startup properties of the EM and can lead to an error in determining the overload capability of the EM due to the inaccurate determination of the maximum slip. Therefore, it is advisable to fully determine the startup characteristics for the entire range of variation of the slip from startup to values corresponding to a mode close to the rated one.

According the specific methodology used – a developed mathematical model has been used to study the transient processes in the electric drive of pump unit with medium voltage IM [12]. The coordinate system, in which the voltages of the stator and rotor windings of IM are presented in relative units, rotates at synchronous speed. Using this coordinate system provides the convenience that in the system of differential equations attend important parameter of the IM *slip s*. Some more important data values and parameters for the considered IM are given in the Appendix. The parameters of the T-shaped replacement circuit of the EM are used in the research, which are determined by the calculation methodology for slip $s=1$.

After converting the equations for voltages of windings and presenting in the form of *Cauchy*, we get four equations for model stator currents, as well as an equation for the electromagnetic torque. We use system of relative units. In this system voltages, currents, power and parameters are expressed in parts of the underlying values of those variables. Rated values of current, voltage, power, torque, speed, resistances serve as basic values.

The mechanical equipment is presented by means of a single-mass model using fundamental relation between torques, so-called *equation of motion*. After transformation for the equation of motion has been obtained [13]:

$$\frac{ds}{dt} = -\frac{p}{J_{\Sigma}\omega_b}(T - T_L), \tag{1}$$

where

- p – pole pair number
- T – electromagnetic torque developed by the motor
- T_L – resisting moment of the driven mechanism
- J_{Σ} – total inertia moment of the of the electric drive (EM together with mechanism driven), reduced to the EM shaft
- ω_b – base angular velocity (equal to the synchronous one – ω_0)

In the equation of motion of the electric drive, the reduced moment of inertia of the system must be inserted. The reduced moment of inertia of a given system is the moment of inertia attributed to one shaft (usually that of the EM chosen as the defining component) which, at the angular velocity of that shaft, will cause the same kinetic energy as in the real system. When calculating the dynamic characteristics of an EM, the moments of inertia of the mechanisms attached to its shaft are also taken into account. Their values are brought to the frequency of rotation of the rotor.

Assuming the mechanical connections between the individual masses to be absolutely rigid (with rigidity coefficient $c = \infty$), the mentioned one-mass dynamic model has been obtained. Since a multi-mass dynamic model of the mechanisms is not always available, and its development is labor-intensive, the results obtained with a single-mass model, for which J_{Σ} has been obtained by the so-called *factor of inertia* (*FI*), are of interest:

$$J_{\Sigma} = FI \times J_M, \tag{2}$$

where

- J_M – EM inertia moment (total of the rotor and the shaft)

As for the moment of inertia of the load reduced to the rotor axis, it can be significantly greater than that of the EM rotor. It should be clear that the higher the total inertia, the slower the acceleration and vice versa. However, it is usually assumed that the total inertia is not likely to be more than twice the EM inertia, and this is certainly the case for most loads [14]. In model studies, a value of $FI = 2$ has been set.

Moment of inertia is analogous to mass in reciprocating motion, i.e. moment of inertia determines the inertial properties of bodies in rotational motion. AC electric motors generally have less mechanical inertia than DC motors due to the absence of a collector and frictional brushes.

The moment of inertia also participates in the so-called *electromechanical time constant*, which characterizes the speed of a given EM in the process of its rotation [15].

With regard to the resisting moment of the centrifugal pump as a driven mechanism – it depends on the magnitude of the angular velocity to a certain degree, and the characteristic is parabolic and is represented by a general expression, the so-called *Blanc equation*:

$$T_L = T_{L0} + \left(T_{Lrated} - T_{L0}\right)\left(\frac{\omega}{\omega_{rated}}\right)^k, \tag{3}$$

where

- T_L – resisting moment of the mechanism at speed ω
- T_{L0} – resisting moment of the mechanism at speed $\omega = 0$ (initial moment of resistance)
- T_{Lrated} – rated resisting moment at speed $\omega = \omega_{rated}$
- k – power indicator characterizing the degree of change of the resisting moment from the speed, k=2÷6

For powerful IMs losses at rated load and rated speed is allocated to the rated power as: losses in the stator steel – from 1.95 to 2%; electrical losses in the stator windings – from 0.9 to 1.1%; electrical losses in the rotor windings – from 0.8 to 0.9%; mechanical losses – from 1.3 to 1.5%; additional losses – from 0.5 to 1%, [14, 16].

Electrical losses in the stator and rotor windings are variable losses, since their magnitude depends on the values of the currents in these windings [12, 17]. Electrical losses in the rotor are directly proportional to slip.

The variable electrical losses in the stator and rotor can be given as follows [18]:

$$\Delta P_V = \Delta P_1 + \Delta P_2 \approx 3I_1^2 r_1 + 3I_2'^2 r_2', \tag{4}$$

where

- ΔP_1 – stator electrical losses; ΔP_2 – rotor electrical losses
- I_1 – stator current; I_2' – rotor current reduced to the stator
- r_1, r_2' – stator and reffered rotor phase resistance, respectively

When operating IM in the area of small slips (not more than the rated value), it follows

$$I_1^2 \approx I_2'^2 + I_0^2, \tag{5}$$

where:

- I_1, I_2', I_0 – stator, rotor and magnetizing current, respectively

Therefore, the following applies to IM:

$$\Delta P_V = 3I_2'^2\left(r_1 + r_2'\right). \tag{6}$$

Since

 https://doi.org/10.1051/e3sconf/202340403005

$$I_2' = \sqrt{\frac{T\omega_0 s}{3r_2'}}, \tag{7}$$

then the representation of the total electrical losses takes the form

$$\Delta P_V = \Delta P_1 + \Delta P_2 = T\omega_0 s(1 + \frac{r_1}{r_2'}). \tag{8}$$

The starting time is the main quantity that determines the heating of the EM during the starting process – the longer this time, the harder the starting process and the higher the EM heating temperature will be. This time is determined by the criterion that two adjacent values of the rotation frequency do not differ by more than 1%. This means that the electromechanical transient processes are complete, and the electromagnetic transient processes are known to terminate much earlier.

3 Simulation results

Appropriate software MathCAD has been used to solve the system of differential equations describing the transient processes occurring during the start-up of a MV induction EM driven pump unit [19]. The analysis of the transient processes has been carried out under the generally accepted assumptions, which do not have a significant impact on the final results, and on the other hand, make it easier to understand the physical nature of the processes studied. The start-up of the electric drives is accompanied by a redistribution of the consumed power between the output power and the losses. Of greatest interest are the losses in the rotor, which, with frequent starts, can lead to damage to its windings. At the same time, in their determination through a definite integral for the release time, the moment of inertia is involved as a quantity.

The results of the model studies on the influence of the control way applied on the impact stator currents and torques of the EM, as well as on the electrical power losses and startup time are systematized in Table 1 and Table 2. For a better understanding, some of the dependencies obtained have been presented graphically in Figure 1 ÷ Figure 9. In order to establish the relevant trends corresponding polynomial dependencies have been derived.

Table 1. Results for softstarter applied at changing the voltage from 30% to 100% of rated one.

Quantities				
U/U_{rated}, %	T^*_{imp}	i^*_{imp}	$\Delta P_{electric}$, x10^4 W	t_{start}, s
30	2.088	4.964	4.410	2.903
40	3.709	6.805	2.398	1.912
50	5.790	8.397	1.519	1.151
60	8.328	10.445	1.051	0.826
70	11.319	12.366	0.771	0.623
80	14.761	13.859	0.589	0.477
90	18.646	15.106	0.466	0.435
100	22.970	16.331	0.377	0.358

All quantities considered are modified in a different way, which is explained under each respective figure.

Table 2. Results for control applied by law U/f^2 = const.

Quantities					
f, Hz	*U*, V	T^*_{imp}	i^*_{imp}	$\Delta P_{electric}$, x10^4 W	t_{start}, s
20	960	1.481	2.715	6.532	2.798
25	1500	2.889	4.381	3.107	1.377
30	2160	4.985	6.286	1.764	0.834
35	2940	7.907	8.369	1.103	0.652
40	3840	11.788	10.852	0.742	0.543
45	4860	16.705	13.454	0.518	0.431
50	6000	22.970	16.331	0.377	0.358
55	6000	20.940	17.128	0.419	0.423
60	6000	19.231	17.526	0.446	0.427

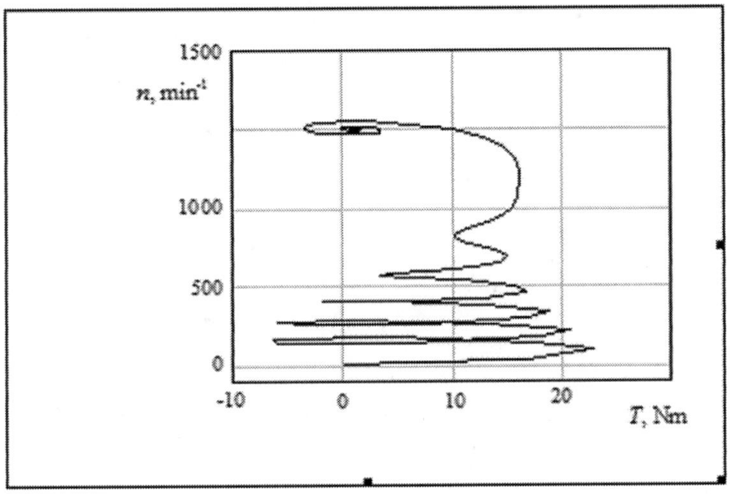

Fig. 1. Torque-speed characteristic for DOL starting.

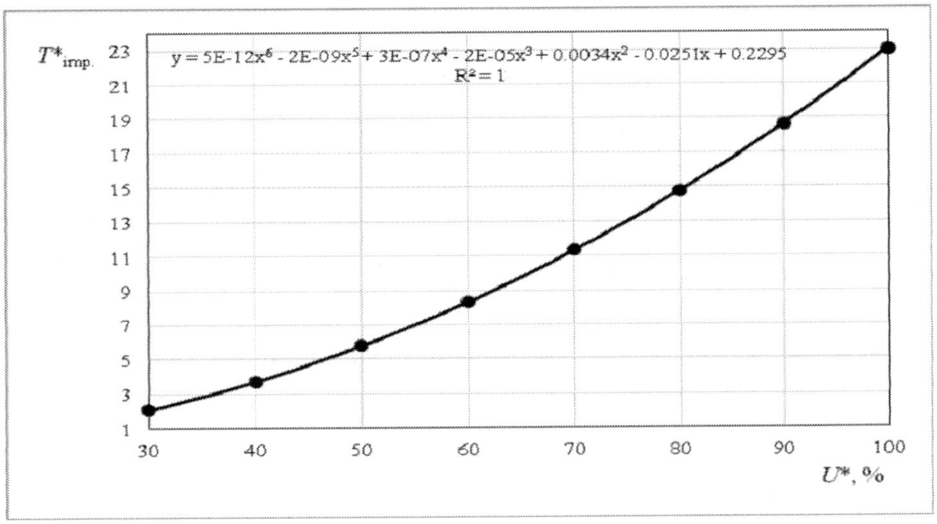

Fig. 2. Variation of the impact torque relative to the supply voltage during soft start.

The arising impact torque can be reduced over 10 times compared to DOL-starting. It is known that motor torque capability is proportional to the square of the applied voltage.

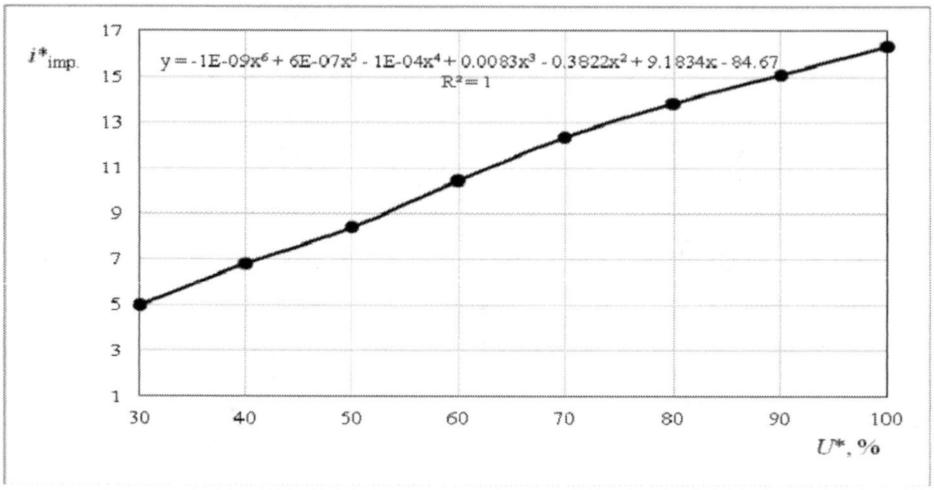

Fig. 3. Variation of the impact current relative to the supply voltage during soft start.

The resulting impact stator current can be reduced more than 3 times, compared to DOL-starting.

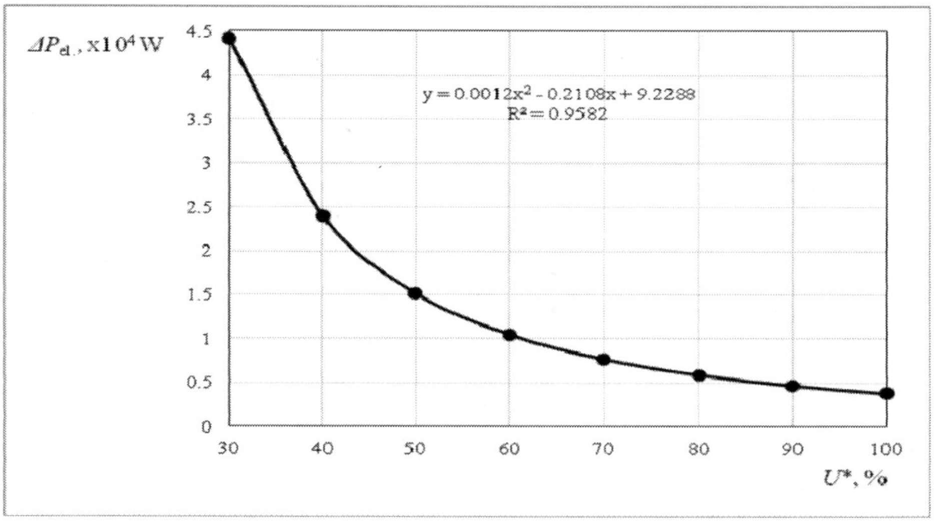

Fig. 4. Variation of electrical power losses relative to supply voltage during soft start.

Electrical power losses increase significantly with each decrease in the value of the supply voltage.

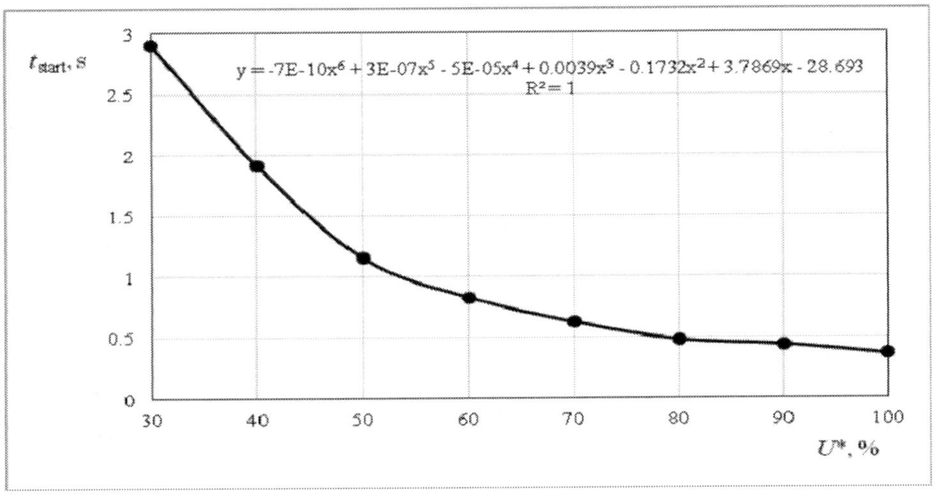

Fig. 5. Variation of starting time relative to supply voltage during soft start.

The start-up time is noticeably increased with each decrease in the value of the supply voltage.

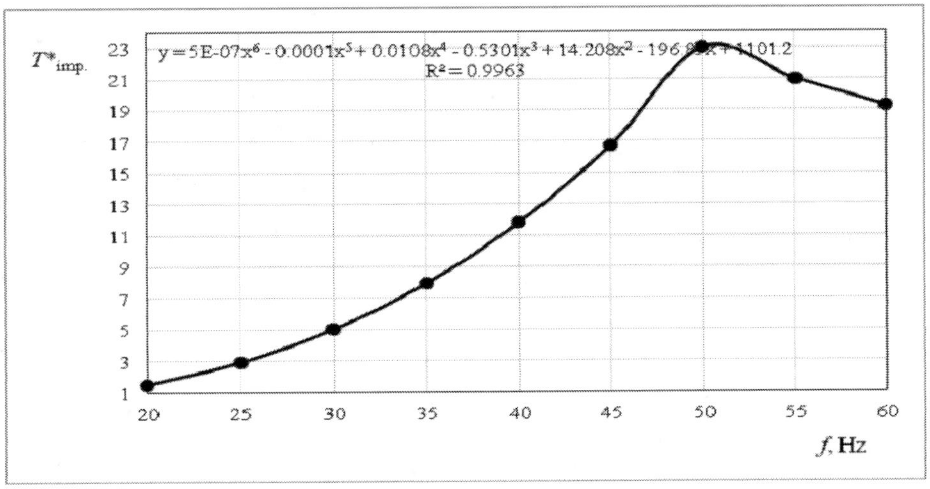

Fig. 6. Variation of the impact torque relative to the frequency of the supply voltage during control U/f^2 = const.

The resulting impact torque has a peak value at the rated frequency of the supply voltage value. At over-synchronous frequencies, its reduction begins, which is positive and could be applied in some cases.

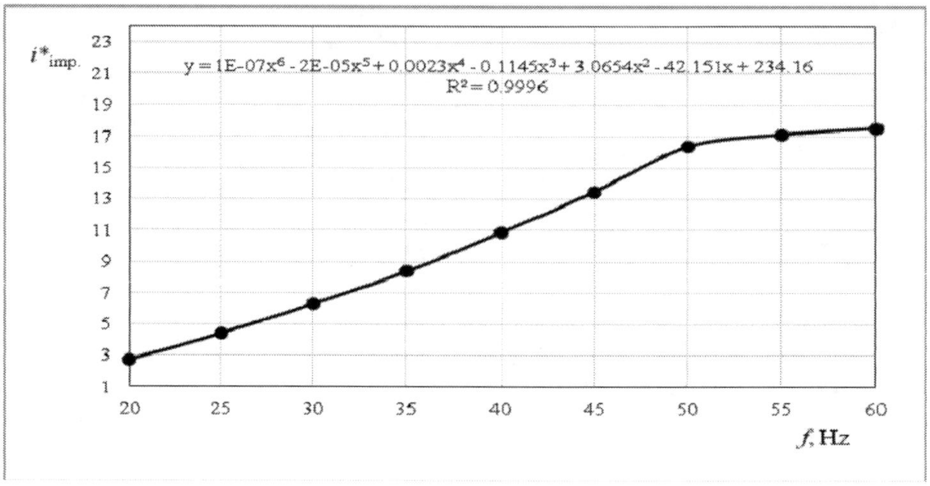

Fig. 7. Variation of the impact current relative to the frequency of the supply voltage during control U/f^2 = const.

The resulting impact stator current can be reduced over 6 times compared to DOL-starting.

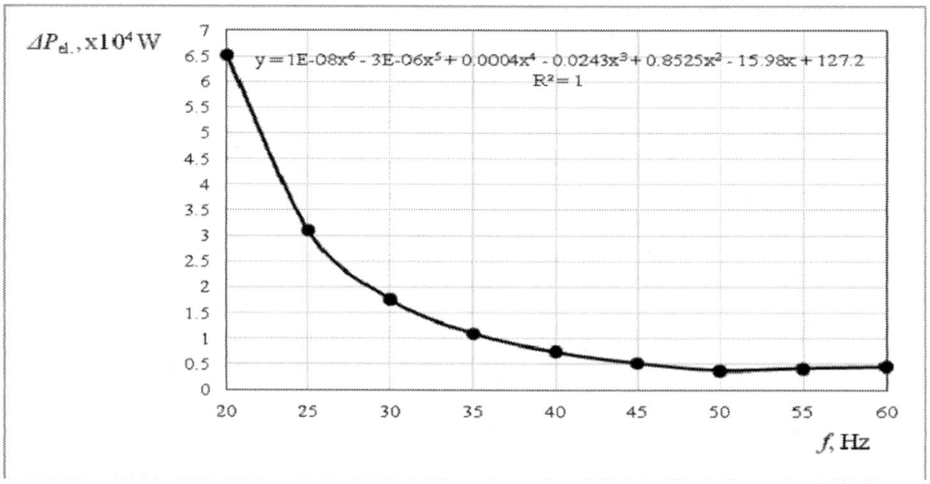

Fig. 8. Variation of electrical power losses relative to the frequency of the supply voltage during control U/f^2 = const.

The change in electrical power losses with frequency control is similar to that with the application of a softstarter.

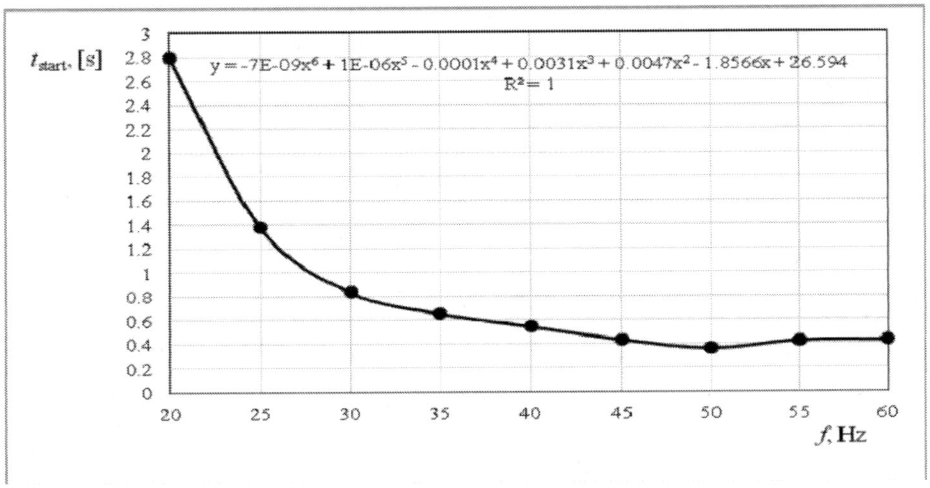

Fig. 9. Variation of starting time relative to the frequency of the supply voltage during control U/f^2 = const.

The starting time has a minimum value at rated voltage and frequency.

4 Conclusion

The application of softstarter and frequency control gives good results in terms of reduction of impact torques and currents. The specifics in the given case are:

- the large values of electrical power losses are at higher slip values in the interval 0÷1: in this case it is about the start-up process, in established mode these losses are significantly smaller, on the order of up to 2%
- starting times are also affected by the reduced voltage level in both modes of control.

Speed control by changing the frequency is the most acceptable option for IMs, as it provides speed control over a wide range without significant losses and reducing the motor's overload capabilities.

The obtained results could be of great practical importance when considering a way of starting and possibly controlling the speed of IM.

Measurements of electrical quantities of the specific IM under consideration are foreseen, which implies providing a measurement technique with increased accuracy and a certain period of time.

As a summary – there is a clear interest in cleaner, greener energy, and hence also for such a world. This implies actions to support energy optimization, environmental protection and resource saving both at the design stage and in operation of electric drive systems.

The implementation of this paper is thanks to the support from:

- the Project BG05M2OP001-1.002-0023 Competence Center 'Intelligent Mechatronic, Eco- and Energy-Saving Systems and Technologies' financed by the Operational Program 'Science and Education for Intelligent Growth' of the European Regional Development Fund.
- the University Project 2206S 'Applied mathematical investigations in electric power supply and electrical equipment focused on energy and economic efficiency', financed by the Ministry of Education and Science of the Republic of Bulgaria.

References

1. B. Stoffel, Assessing the energy efficiency of pumps and pump units, Elsevier, Ltd., Waltham, U.S.A., pp. 3-4 (2015)

2. J. A. Kay, R. H. Paes, J. G. Seggewiss, R. G. Ellis, IEEE Transactions on Industry Applications, **36**(6), pp. 1688-1689 (2000)
3. EN 50160 Voltage characteristics of electricity supplied by public electricity networks, (2010)
4. Pichai Aree, *Effects of small and large induction motors on voltage sag profile*, ECTI-CON, Phetchaburi, pp. 1-4 (2012)
5. P. Krause, O. Wasynczuk, S. Sudhoff, S. Pekarek, Analysis of electric machinery and drive systems, (John Wiley & Sons, Inc., Hoboken, New York, U.S.A. 2013)
6. J.A. Melkebeek, Electrical Machines and Drives: Fundamentals and Advanced Modelling, (Springer International Publishing AG, Switzerland, 2018)
7. F.D. Petruzella, Electric motors and control systems, (McGraw-Hill, New York, U.S.A. 2016)
8. P. Wach, Dynamics and Control of Electrical Drives, (Springer-Verlag Berlin Heidelberg, 2011)
9. S.S. Rao, Engineering Optimization: Theory and Practice, 3rd edition, (John Wiley & Sons, New York, U.S.A., 1996)
10. I. Boldea, L. Tutelea, Electric machines: steady state, transients, and design with MATLAB, (CRC Press, Boca Raton 2009)
11. V.S. Lobanoff, R.R. Ross, Centrifugal pumps: design & application, 2nd edition, (Gulf Publishing Company, 1992)
12. S.Rachev, K. Dimitrova, D. Koeva, L. Dimitrov, The Journal of CIEES, **1**(1), pp. 40–49, (2021), https://doi.org/10.48149/jciees.2021.1.1.7
13. S. J. Chapman, Electric Machinery Fundamentals, (McGraw-Hill, New York, U.S.A., 1985)
14. A. Hughes, Electric Motors and Drives. Fundamentals, Types and Applications, 3rd edition, (Elsevier Ltd. Oxford, Great Britain, 2006)
15. R. Crowder, Electric Drives and Electromechanical Systems, (Elsevier, Oxford, U.K., 2006)
16. G. McPherson, R.D. Laramore, An Introduction to Electrical Machines and Transformers, (John Wiley & Sons, New York, U.S.A., 1990)
17. S. Rachev, D. Koeva, I.D. Ivanov, *Electromechanical Model and Operating Modes of High-voltage Induction Motor Electric Drive – Part I: Impact Torque and Currents*, in Proceedings of the 2016 International Symposium on Fundamentals of Electrical Engineering (ISFEE), Bucharest, Romania, pp. 1-5 (2016)
18. H. Merz, Electrical Machines and Drives. (Berlin, Germany: VDE VERLAG GmbH, 2002)
19. PTC Mathcad, Parametric Technology Corporation (2020)

https://doi.org/10.1051/e3sconf/202340403005

Appendix

Technical data and parameters for induction motor considered

Quantity	Value
Type	A 121-4
Rated power	400 kW
Rated stator voltage	6000 V
Operating frequency	50 Hz
Line stator current	46.5 A
Pole pair number	2
Shaft speed	1480 rpm
Power factor	0.89
Efficiency	93.1%
Rotor moment of inertia	22.5 kg.m2
Multiple of the maximum torque	2.1
Multiple of the maximum current	5.1
Stator phase resistance	1.574 Ω
Referred rotor resistance	0.811 Ω
Stator leakage reactance	4.257 Ω
Referred rotor leakage reactance	0.662 Ω
Magnetizing reactance (saturated)	157.954 Ω

Reduction of earthquake amplification by seismic metamaterials made of circular positioned concrete piles

Selcuk Kacin[1], *Murat* Ozturk[1], *Umur* Sevim[1], *Muharrem* Karaaslan[2*], *Oguzhan* Akgol[2], *Zafer* Ozer[3], *Mustafa* Demirci[1], *Emin* Unal[4], *Bayram* Mert[5], *Mustafa* Başar[6]

[1]Iskenderun Technical University, Dept. of Civil Eng., Hatay, 31200, Turkey, 31200, Turkey
[2]Iskenderun Technical University, Dept. of Electrical and Electronics Eng., Hatay, 31200, Turkey
[3]Mersin University, Vocational School of Mersin, Dept. of Electronics Tech., Mersin, 33350,
[4]Necmettin Erbakan University, Electrical and Electronics, Konya, 42090,
[5]Iskenderun Technical University, Iskenderun Vocational School, Hatay, 31200, Turkey
[6]Iskenderun Technical University, Dept. of Mechanical Eng., Hatay, 31200, Turkey

Abstract. Seismic surface waves with low frequencies results in higher effects. The aim of the utilization of seismic metamaterials is to prevent or minimize the effects of this waves particularly wide frequency ranges. This study presents the investigation of the reducing effects of seismic metamaterials composed of circular array concrete piles on surface waves. The investigation of circular array concrete piles has been carried out numerically between 5-15 Hz. The effectiveness of the periodic structure has been observed numerically for fields including metamaterials by comparing empty field. The observations have been depicted according to transmission losses for two cases by using finite element method (FEM) simulations. The wave propagation also presented at the related frequencies where the transmission losses seen. Effective dimensions of radius and length of the structures have been obtained by using parametric approach. It has been proved numerically that the suggested concrete piles reduces the hazardous effects of surface waves.

1 Introduction

Seismic surface waves with low frequencies results in higher effects [1, 2, 3]. The aim of the utilization of seismic metamaterials is to prevent or minimize the effects of this waves particularly wide frequency ranges. This study presents the investigation of the reducing effects of seismic metamaterials composed of circular array concrete piles on surface waves. The investigation of circular array concrete piles has been carried out numerically between 5-15 Hz. The effectiveness of the periodic structure has been observed numerically for fields including metamaterials by comparing empty field.
The observations have been depicted according to transmission losses for two cases by using finite element method (FEM) simulations [4, 5]. The wave propagation also presented

* Corresponding author: muharrem.karaaslan@iste.edu.tr

at the related frequencies where the transmission losses seen [6]. Effective dimensions of radius and length of the structures have been obtained by using parametric approach [7]. It has been proved numerically that the suggested concrete piles reduces the hazardous effects of surface waves [8].

Seismic metamaterials have emerged as a novel approach to mitigate low-frequency waves within the resonance frequency range of 1-10 Hz [9]. Inspired by the vibration damping properties of forests, Liu et al. [10] utilized trees as large-scale natural metamaterials to dampen low-frequency vibrations. Miniaci et al. [11] explored the potential of large-scale mechanical metamaterials for passive isolation, impacting both surface and guided waves. Pu and Shi [12] investigated the behavior of surface waves in one- and two-dimensional periodic structures, demonstrating the effectiveness of a finite periodic pile system in attenuating waves.

This study investigates the effectiveness of seismic metamaterials formed by placing concrete piles in a circular array to prevent surface waves. The proposed structure was analysed theoretically using a finite element method (FEM)-based analysis program. The study was conducted in both the frequency and time domains, with a focus on the low-frequency range, which poses a threat to structures. The proposed structure shows promise as an effective method for mitigating the propagation of seismic surface waves and could be considered for use in earthquake research.

2 Theory and numerical analysis

In this study, the numerical analysis has been carried out by using finite element method (FEM) based simulator. The circular arrangement of concrete piles based seismic metamaterials has been assigned in the field of simulation as can be seen in Figure 1.

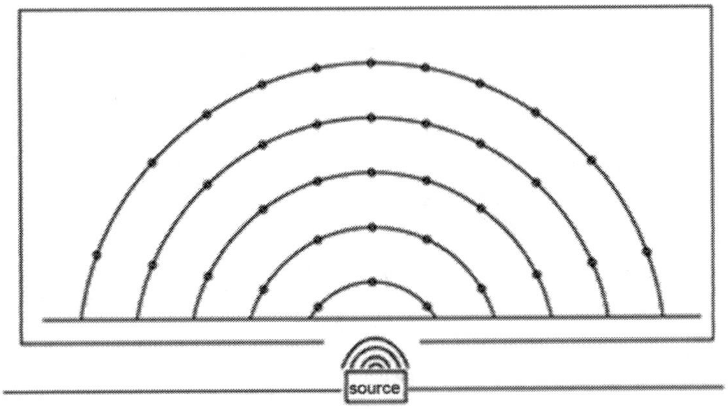

Fig. 1. Concrete piles in circular arrangement.

The length and radius of the concrete piles has been evaluated by using parametric approach. The effective values of the radius and length of the concrete piles have been determined as 7.5cm and 2m, respectively. Since the non-periodical arrangement of the piles, the band diagram of one unit cell cannot be extracted. The Young's modulus of the soil is assigned as 20 MPa, the poisson's ratio defined as 0.3 and the density is denoted as 1800 kg / m3, the Young modulus, the poisson's ratio and the density are assigned as 30 GPa, 0.25 and 2500 kg / m3, respectively. All these values are obtained from the field analysis and measurement results of the soil. The observation of the effectiveness of the proposed concrete piles has been realized by comparing the results of soil with and without concrete piles in circular array. The overall system has been depicted in Figure 2.

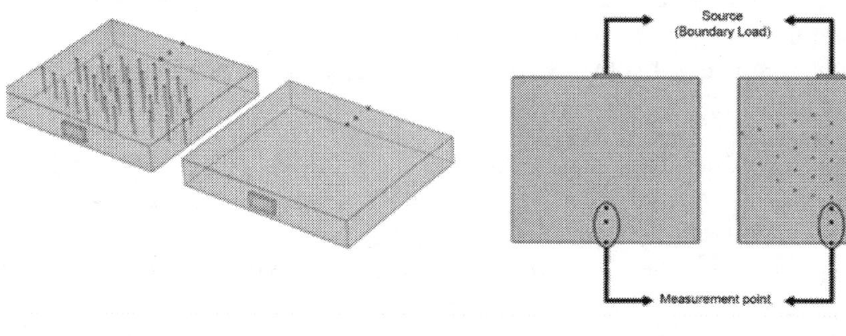

a) b)

Fig. 2. a) Fields with and without concrete piles b) measurement points and boundary loads.

1 N test signal has been applied to sample fields in simulations for both frequency and time domain cases. The frequency range has been defined between 5-15 Hz which is highly destructive characteristics. The applied signals have been depicted in Figure 3 in frequency and time domain.

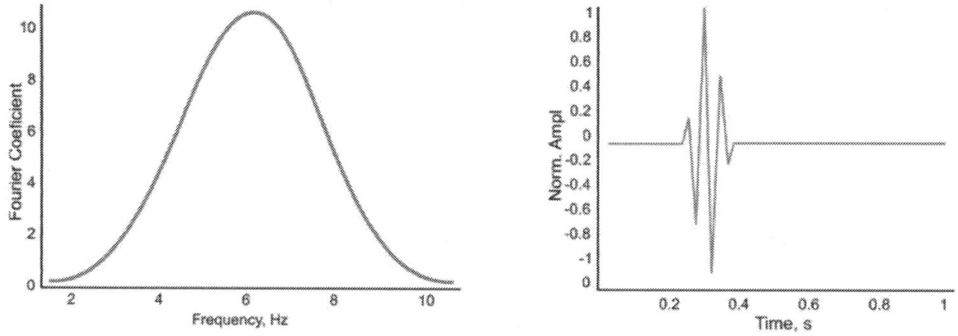

Fig. 3. Applied pulses in frequency and time domain.

The transmission losses in terms of dB for concrete piles with respect to empty field have been demonstrated in Figure 4. Six resonance frequency points have been observed between 5-15 Hz. The resonance frequencies and transmission loses for them are 5.1 Hz, 6.2 Hz, 6.5 Hz, 8.5 Hz, 11.2 Hz, 12.3 Hz, 13.8 Hz and 14 Hz, -16 dB, -11.73 dB, -20.6 dB, -13.3 dB, -23.28 dB, -28.1 dB, -18.85 dB, -37.82 dB, respectively.

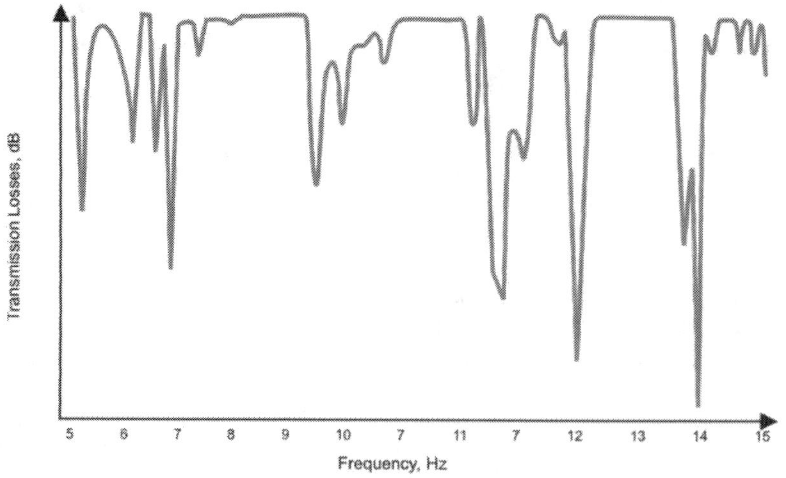

Fig. 4. Transmission Values in dB.

The propagation of the applied vibration in the field with and without metamaterials has been given in Figure 5 for each resonance frequencies mentioned above. Hence, it can be possible to compare the effectiveness of the concrete pills and circular arrangement.

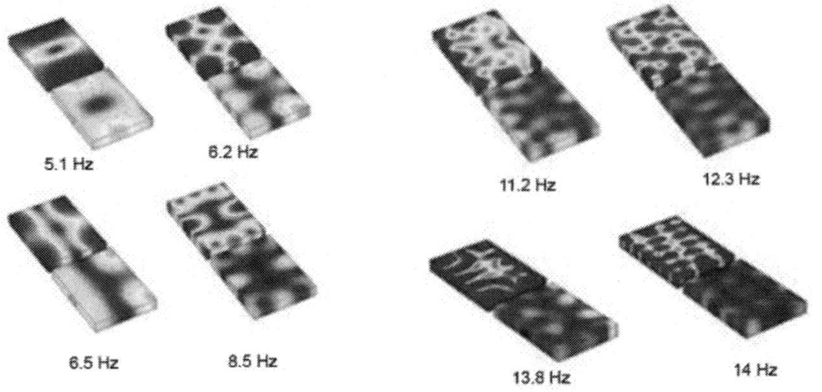

Fig. 5. Total displacement with and without boreholes at resonance frequencies.

The simulations have been realized for each resonance frequency, since the transmission coefficients are not equal for different cases. The maximum and minimum displacements scales have been demonstrated by red and blue colours. As can be seen from the results, the propagation of vibration reduces behind the metamaterial structures.

The simulations have also been carried out in time domain at the related resonance frequencies by using finite element method based simulator. Since the surface waves propagate perpendicular to concretes, the reduction characteristics of waves investigated in x-y plane. Besides, the horizontal component of the applied waves along x direction resonates with concrete piles. The phase difference between incident and reflected vibrations results in reduction of propagating wave in resonance case. The responses of the periodic structure against surface wave for three discrete frequencies have been depicted in Figure 6.

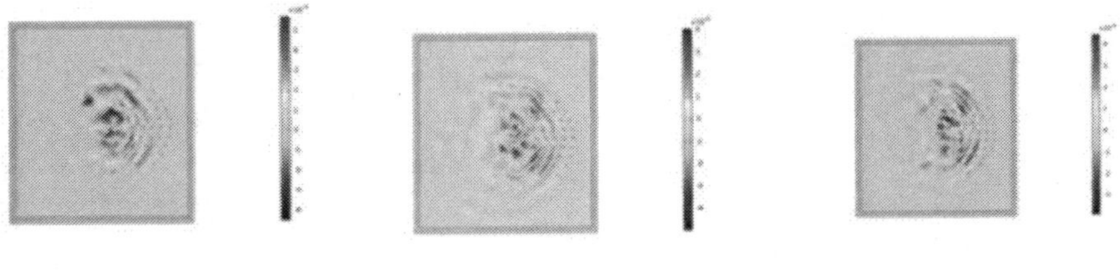

a) b) c)

Fig. 6. a) 2D displacement at 6.5 Hz b) 2D displacement at 8.5 Hz c) 2D displacement at 14 Hz.

Upon analysing wave propagation at specific frequencies within the field incorporating concrete piles, it has been observed that the waves experienced damping after passing through three rows of holes. The wave gradually weakened as it propagates and ultimately failed to pass beyond the proposed structure. The numerical studies conclusively determined that the seismic metamaterial formed by concrete piles arranged in a circular array effectively decreases seismic waves.

3 Conclusion

In this study, circularly arranged concrete piles based seismic metamaterials has been proposed to confine the propagation of vibrations. The simulations have been carried out by using finite element based simulator in the frequency range of 5-15Hz. This frequency range has been chosen due to strong destructive properties. The optimum dimensions of the piles have been determined by using parametric approach. The analysis has been realized for field including concrete piles and the transmission results have been compared with field without piles. The simulation results demonstrated that the periodic structure has capability to reduce propagating wave at the frequencies of 6.5 Hz, 8.5 Hz and 14 Hz. The transmission losses for the resonance frequencies of 6.5 Hz, 8.5 Hz, and 14 Hz are -20.6 dB, -18.85 dB, -37.82 dB, respectively. Hence, it can be concluded that the proposed structure has capability to prevent seismic wave propagation efficiently especially at the resonance frequencies.

This research was supported by the Disaster and Emergency Management Presidency of Turkey (UDAP-Ç-19-21) for the financial support and presentation support by 2224-A Grant Program for Participation in Scientific Meetings Abroad with application number of 1919B022301764.

References

1. D. W. Prather, S. Shi, J. Murakowski, G. J. Schneider, A. Sharkawy, C. Chen, ... & R. Martin, *Self-collimation in photonic crystal structures: a new paradigm for applications and device development*. Journal of Physics D: Applied Physics, 40(9), 2635 (2007)
2. R. Martínez-Sala, J. Sancho, J. V. S´anchez, V. G´omez, J. Llinares, F. Meseguer. *Sound attenuation by sculpture*. Nature;378:241 (1995)
3. R. Zhu, X. N. Liu, G. K. Hu, C. T. Sun, & G. L. Huang, *Negative refraction of elastic waves at the deep-subwavelength scale in a single-phase metamaterial*. Nature communications, 5(1), 1-8 (2014)
4. J. Wen, H. Shen, D. Yu, & X. Wen, *Exploration of amphoteric and negative refraction imaging of acoustic sources via active metamaterials*. Physics Letters A, 377(34-36), 2199-2206 (2013)
5. S. D. Zhao, Y. S. Wang, & C. Zhang, *Acoustic imaging and mirage effects with high transmittance in a periodically perforated metal slab*. Journal of Applied Physics, 120(19), 194901 (2016)
6. S. Gonella, A. C. To, & W. K. Liu, *Interplay between phononic bandgaps and piezoelectric microstructures for energy harvesting*. Journal of the Mechanics and Physics of Solids, 57(3), 621-633 (2009)
7. K. Zhang, C. Zhao, P. Zhao, J. Luo, & Z. Deng, *Wave propagation properties of rotationally symmetric lattices with curved beams*. The Journal of the Acoustical Society of America, 148(3), 1567-1584 (2020)
8. M. Gao, & Z. Shi, *A wave guided barrier to isolate antiplane elastic waves*. Journal of Sound and Vibration, 443, 155-166 (2019)
9. S. Brûlé, E. H. Javelaud, S. Enoch, & S. Guenneau, *Experiments on seismic metamaterials: molding surface waves*. Physical review letters, 112(13), 133901 (2014)
10. Y. F. Liu, J. K. Huang, Y. G. Li, & Z. F. Shi, *Trees as large-scale natural metamaterials for low-frequency vibration reduction*. Construction and Building Materials, 199, 737-745 (2019)
11. M. Miniaci, A. Krushynska, F. Bosia, & N. M. Pugno, *Large scale mechanical metamaterials as seismic shields*. New Journal of Physics, 18(8), 083041 (2016)

https://doi.org/10.1051/e3sconf/202340403006

12. X. Pu, & Z. Shi, *A novel method for identifying surface waves in periodic structures*. Soil Dynamics and Earthquake Engineering, 98, 67-71 (2017)

AUTHOR INDEX